Doodle
yourself smart...
Math
book
H
I
A
C
L
E
A
B
a
b
a
y=f(x)

$x^2 + 2xy - 2y^2 + x = 2$

$\frac{d(x^2)}{dx} + \frac{d(2xy)}{dx} - \frac{d(2y^2)}{dx} + \frac{d(x)}{dx}$

$2x + (2y + 2x\frac{dy}{dx}) - 4y\frac{dy}{dx} + 1$

$\frac{dy}{dx}(2x - 4y) = -1 - 2x - 2y$

$\frac{dy}{dx} = \frac{-1 - 2x - 2y}{2x - 4y}$

×6

×2

×8

1) $\frac{x}{5} + 2 = 8$ 2) $\frac{w}{3} - 5 = 2$ 3) $\frac{x}{8} + 3 = 12$

$\frac{d(x^3)}{dx} + \frac{d(3y^4)}{dx} - \frac{d(y^2)}{dx} - \frac{d(2x)}{dx} = 0$

$500\ m.s^{-1}$

$9.8t$

v_2

Doodle Yourself Smart...

Doodle yourself smart...
Math book
over 100 doodles and problems to solve!
THUNDER BAY
P·R·E·S·S
San Diego, California

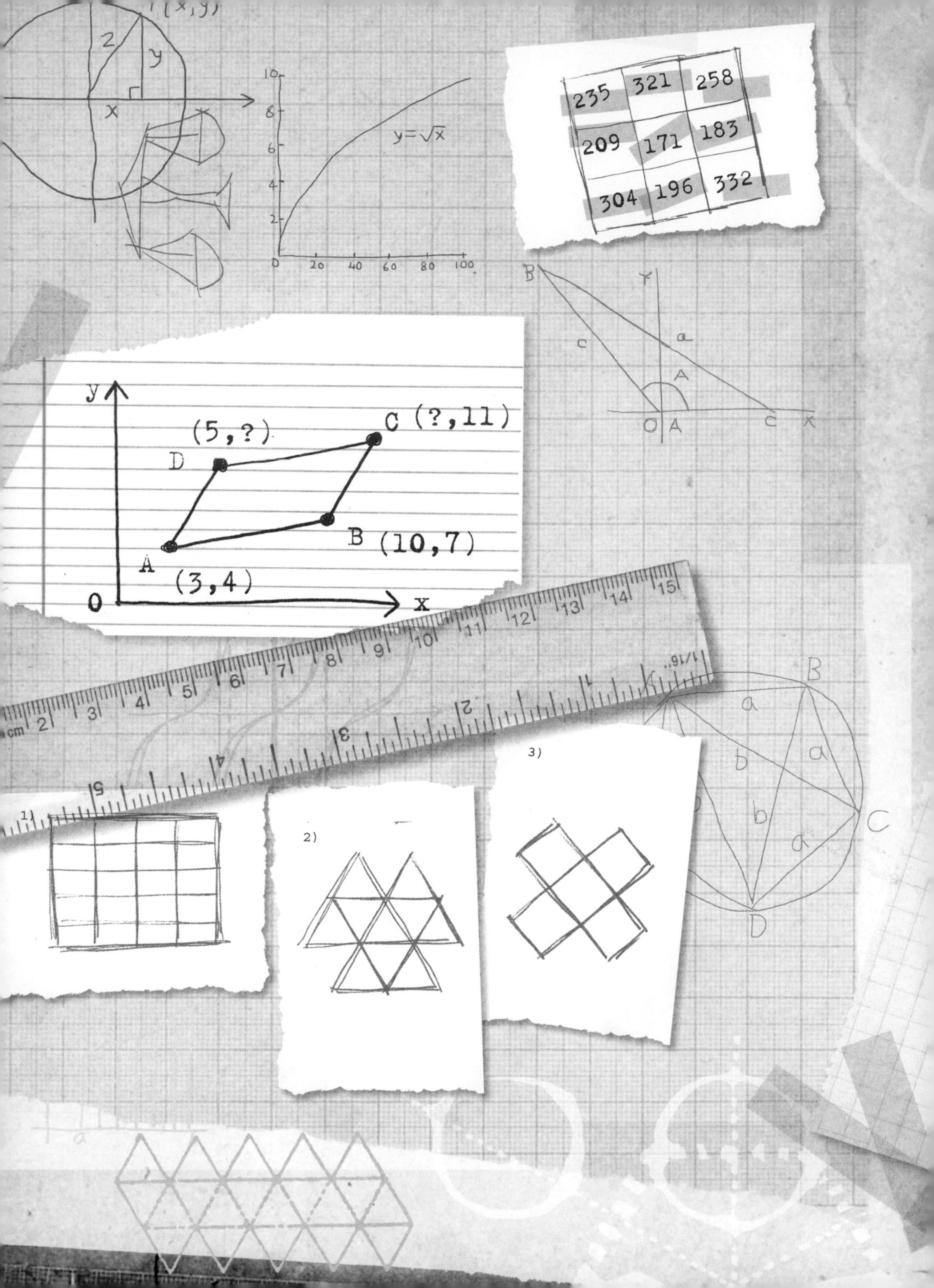

235
321
258
209
171
183
304
196
332
y=√x
(5,?)
C (?,11)
D
B (10,7)
A
(3,4)
1)
2)
3)

y
y = x - 3
x
B
This book belongs to...
Vicky
Doodle Yourself Smart...
G
A
F
B
8
1
6
3
7
4
9
2

Here are four digit cards.

1) Use two of the digit cards to make a multiple of 13. 8 + 5

2) Which two cards could be used to make a factor of 36?

6 x 6

"Pure mathematics is, in its way, the poetry of logical ideas."

[Albert Einstein]

In the number pyramid, the number in each brick is found by adding together the numbers in the two bricks below.

Complete the pyramid.

100

47 53

20 27 26

7 13 14 12

2 5 8 4 6

answers p120/1

Look at this sequence of numbers.

1) What is the rule for the sequence? -2, +5

2) Which two numbers are next in the sequence?

8, 13

answers p120/1

Here is a sorting diagram.

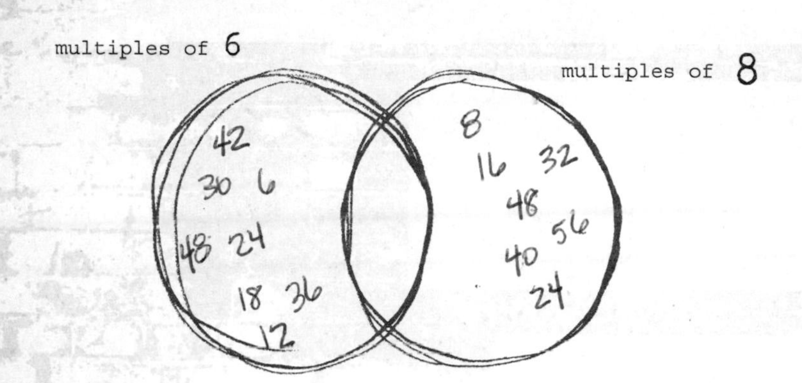

Write each of the numbers below in the correct part of the diagram.

42 8 30 16 6 32 48

40 24 56 18 12 36

1) $\frac{f}{5}+2=8$ 2) $\frac{w}{3}-5=2$ 3) $\frac{x}{8}+3=12$

$d(x^3)$ $d(3v^4)$ $d(v^2)$ $d(2x)$

Fill in the missing numbers to complete the puzzle.

5	x	6	=	30
x		x		
9	x	7	=	63
=		=		
45		42		

answers p120/1

Look at the number grid.

1) Which two numbers in the grid have a total of 454?

2) Which two numbers have a difference of 133?

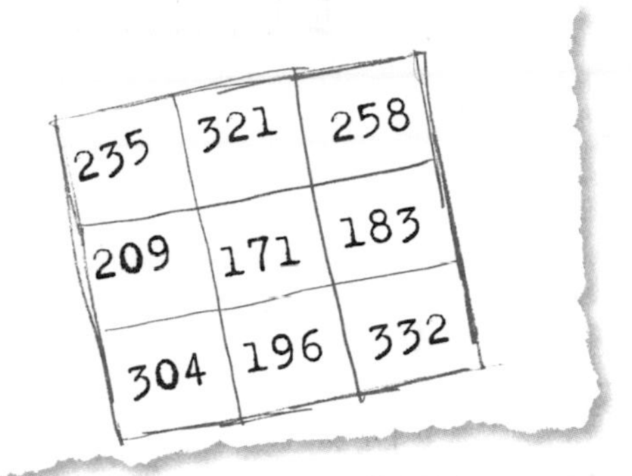

235	321	258
209	171	183
304	196	332

1. 196 + 258

2. 304-171

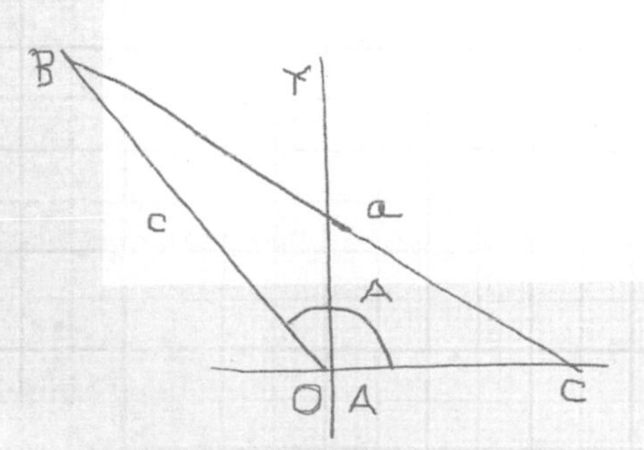

"Mathematics is the only science where one never knows what one is talking about, nor whether what is said is true."

[Bertrand Russell]

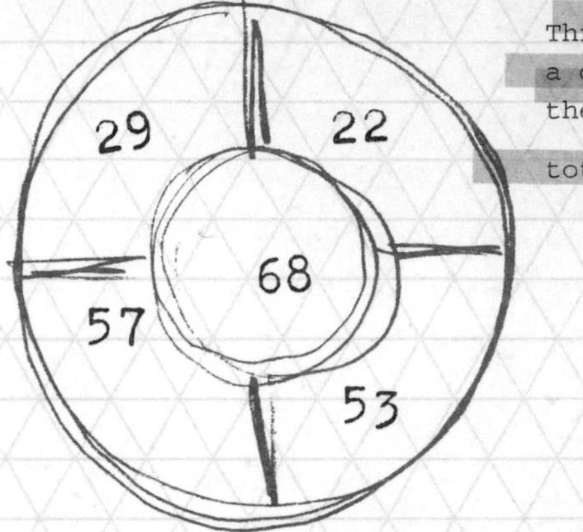

Three darts each land in a different section on the target board. The total score is **150**.

What was the score for each dart?

29
68
53

Complete the multiplication grid.

x	6	7	8	4
4	24	28	32	36
9	54	63	72	36
5	30	35	40	20
3	18	21	24	12

answers p120/1

Here are four digit cards.

1) Make two 2-digit numbers that have the highest total.
2) Make two 2-digit numbers that have the lowest positive difference.

"Mathematics is the queen of sciences and arithmetic is the queen of mathematics."

[Carl Friedrich Gauss]

Which number lies halfway between 1,056 and 2,248?

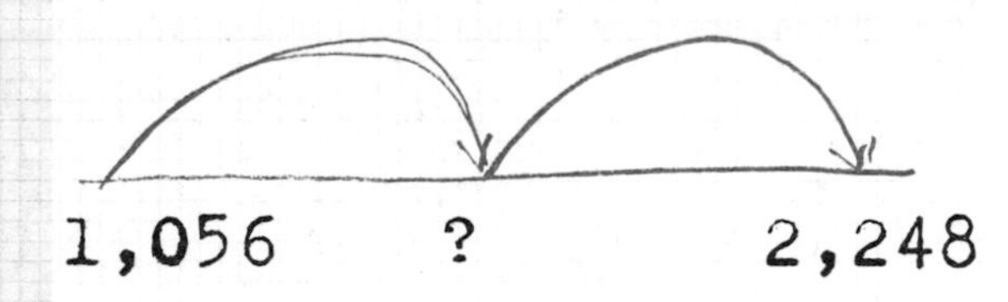

answers p120/1

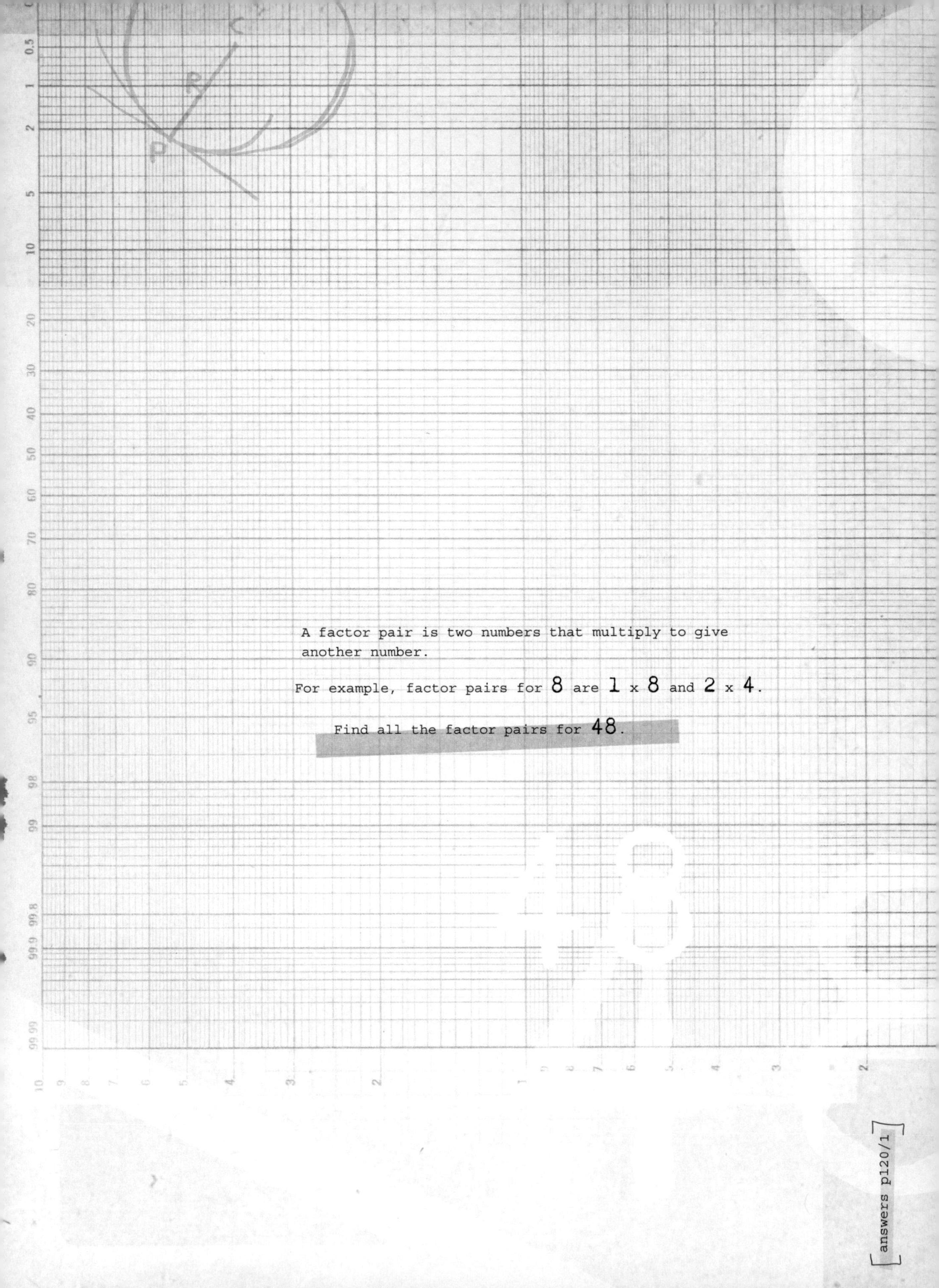

A factor pair is two numbers that multiply to give another number.

For example, factor pairs for 8 are 1 x 8 and 2 x 4.

Find all the factor pairs for 48.

[answers p120/1]

Find a pair of numbers that have a sum of 32 and a product of 252.

$$a + b = 32$$

$$a \times b = 252$$

	e_1	e_2	e_3	e_4
e	e_1	e_2	e_3	e_4
e_2	e_2	$-e_1$	e_4	$-e_3$
e_3	e_3	$-e_1$	$-e_1$	e_2
e_4	e_4	e_2	$-e_2$	$-e_1$

Find the multiple of 45 that is closest to 800.

Complete the magic square so that each column, row, and diagonal adds up to 15.

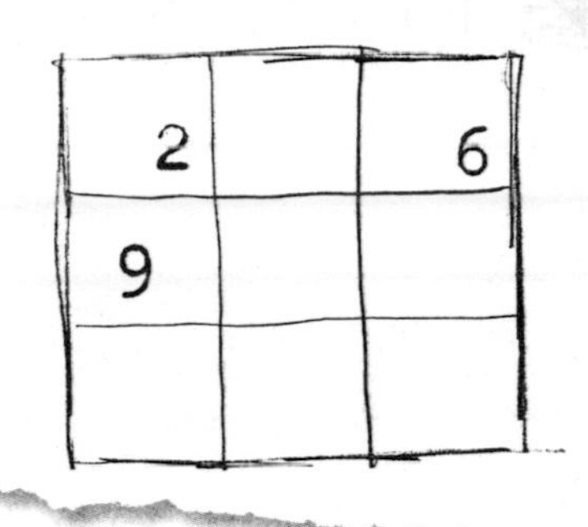

answers p120/1

"The chief forms of beauty are order and symmetry and definiteness, which the mathematical sciences demonstrate in a special degree."

[Aristotle]

Complete these multiplications using only prime numbers.

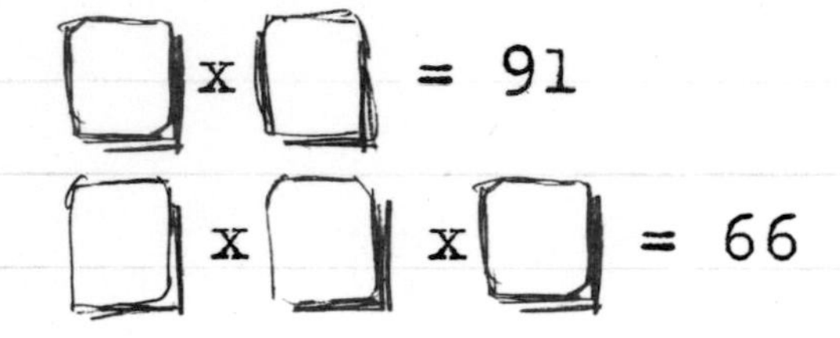

How many seconds are there in two weeks?

	1	2	3	4	5	6
7	8	9	10	11	12	13
14	15	16	17	18	19	20
21	22	23	24	25	26	27
28	29	30	31			

answers p120/1

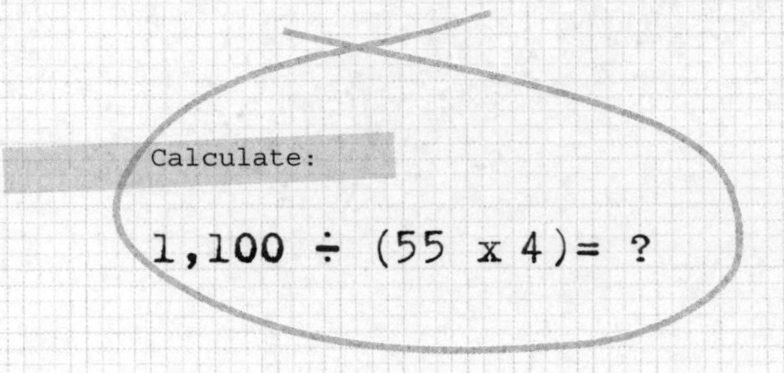

"One can always reason with reason."

[Henri Bergson]

answers p120/1

Here are the charges for a parking lot.

TIME	COST
up to 2 hours	$0.80
up to 3 hours	$1.35
up to 4 hours	$1.90
over 4 hours	$2.30

Car A parks at 09:35 and leaves at 12:10.
Car B parks at 07:50 and leaves at 11:55.

What is the parking charge for each car?

What number is missing from the third triangle?

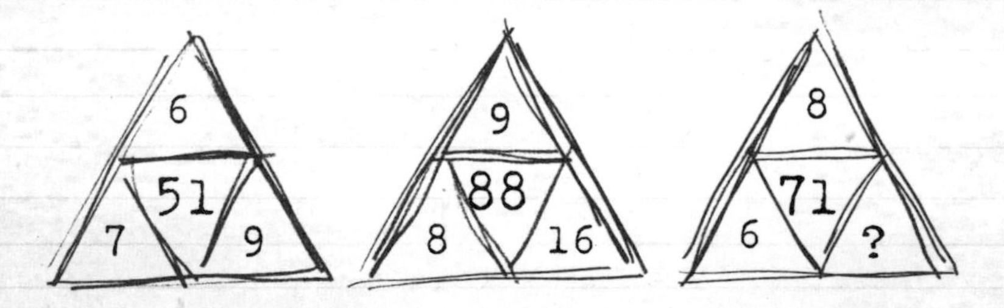

(Hint: look for a link between the numbers in each of the other two triangles.)

Shade all the squares that have a decimal number that is less than 0.25.

0.2	2.5	0.52
0.3	0.15	0.02
0.05	0.03	0.5

answers p120/1

Complete each calculation using the same digit in each box.

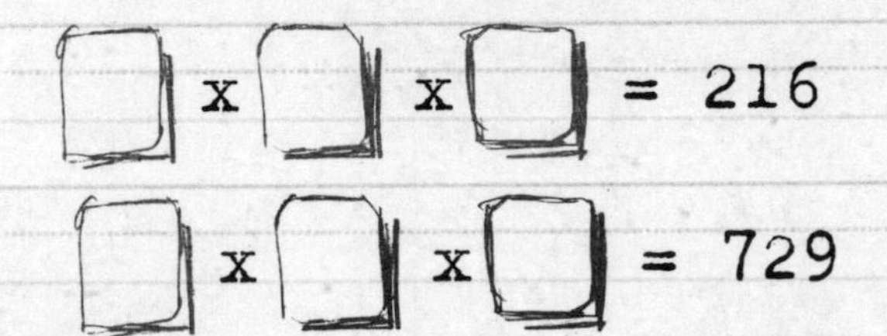

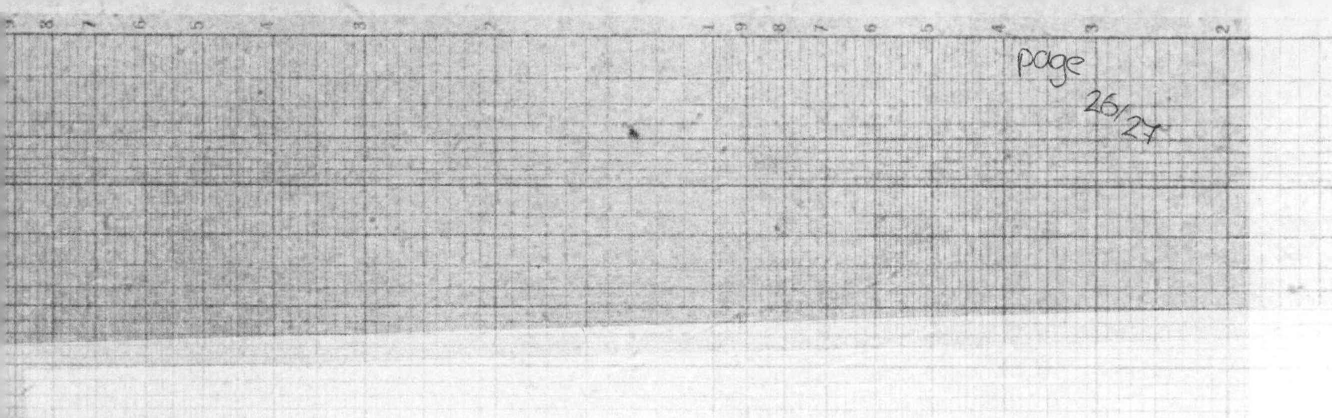

Shade **40%** of each shape.

1)

2)

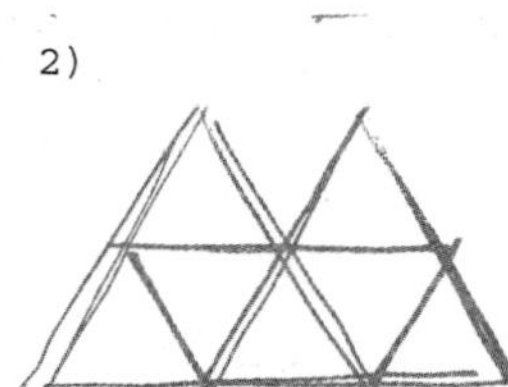

3)

"Perfect numbers, like perfect men, are very rare."

[René Déscartes]

answers p120/1

Circle the fraction, decimal, and percentage that have the same value.

$\frac{3}{5}$ 0.6 $\frac{3}{10}$ 0.35 60% 6% $\frac{10}{16}$

$$\frac{d(x^3)}{dx} + \frac{d(3y^4)}{dx} - \frac{d(y^2)}{dx} - \frac{d(2x)}{dx} = 0$$

page 28/29

Write in the missing digits to make the multiplication correct.

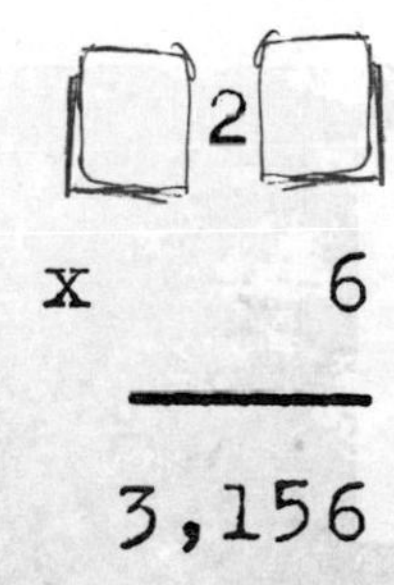

answers p120/1

The three numbers in each row and each column should add up to **109**.

One of the numbers in the grid is incorrect.

Which number is it, and what should it be to make the grid correct?

43	38	28
35	57	15
31	14	64

answers p120/1

150,000 people visited a museum in one year.
15% visited in May and **40%** visited in July.

How many people visited the museum during the rest of the year?

"I have had my results for a long time, but I do not yet know how I am to arrive at them."

[Carl Friedrich Gauss]

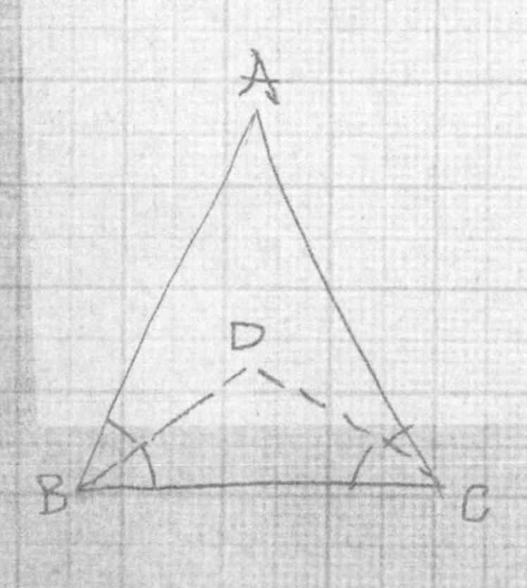

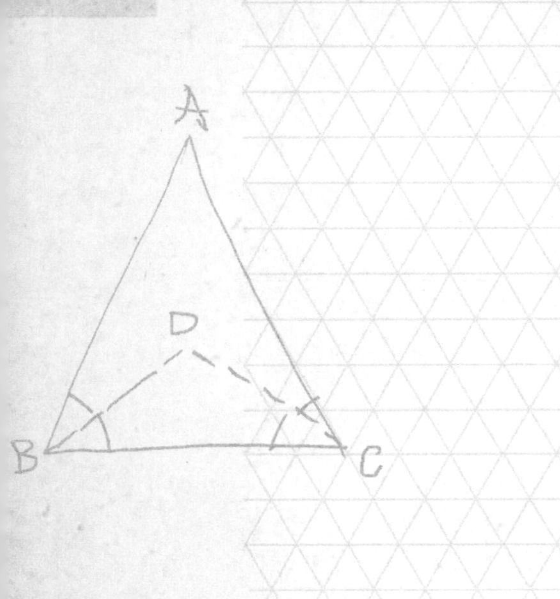

A square number is found by multiplying a number by itself.

For example, 5 x 5 = 25, so 25 is a square number.

Find all the square numbers between 30 and 150.

page 32/33

Here are the ingredients needed to make 8 shortcake cookies

6 ounces of flour
2 ounces of cornstarch
3 ounces of sugar
4 ounces of butter

How much of each ingredient is needed to make 20 cookies?

answers p120/1

"Mathematicians do not study objects, but relations between objects. Thus, they are free to replace some objects by others so long as the relations remain unchanged. Content to them is irrelevant: they are interested in form only."

[Henri Poincaré]

The next number in the sequence is found by doubling the number before it.

Find the missing numbers.

Here is the conversion for distances in kilometers and miles.

8 kilometers = 5 miles

1) What distance in kilometers is 35 miles?

2) What distance in miles is 48 kilometers?

answers p120/1

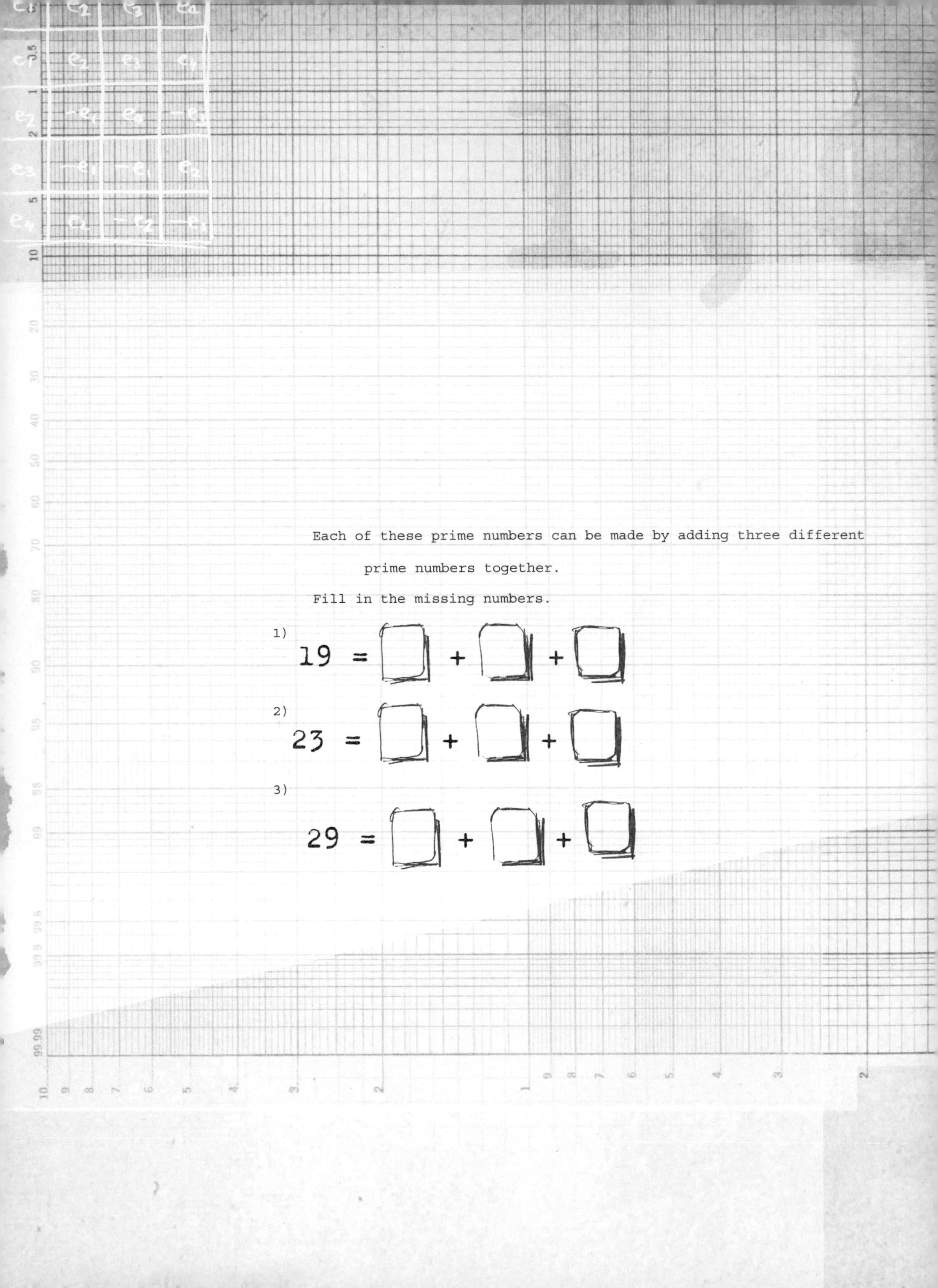

Each of these prime numbers can be made by adding three different prime numbers together.

Fill in the missing numbers.

1) 19 = ☐ + ☐ + ☐

2) 23 = ☐ + ☐ + ☐

3) 29 = ☐ + ☐ + ☐

page 36/37

How much less than 1,000 is $9.9 \times 9.8 \times 9.7$?

answers p122/3

Each number in the sequence decreases by the same amount.

Find the missing numbers.

$36\frac{1}{2}$ □ $25\frac{1}{2}$ □ $14\frac{1}{2}$ □ $3\frac{1}{2}$

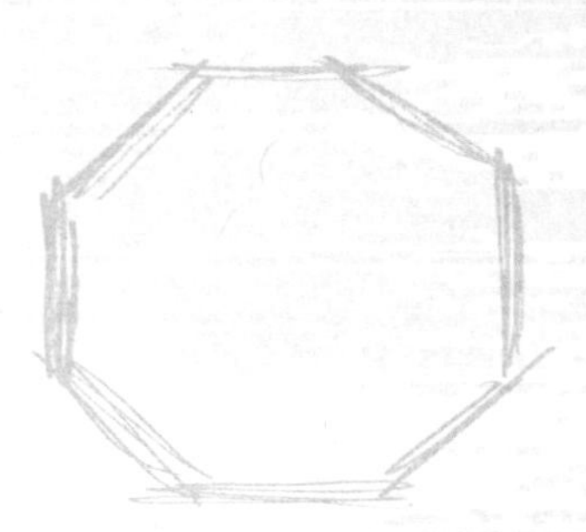

answers p122/3

"In mathematics alone each generation adds a new story to the old structure."

[Hermann Hankel]

A taxi charges a fixed fee of $1.75, plus 50 cents for each mile traveled.

One passenger is charged a fare of $8.25.

How many miles was the taxi journey?

$x^2 + 2xy - 2y^2 + x = 2$

$\frac{d(x^2)}{dx} + \frac{d(2xy)}{dx} - \frac{d(2y^2)}{dx} + \frac{d(x)}{dx} = 2$

$2x + (2y + 2x\frac{dy}{dx}) - 4y\frac{dy}{dx} + 1 = 0$

$\frac{dy}{dx}(2x - 4y) = -1 - 2x - 2y$

$\frac{dy}{dx} = \frac{-1 - 2x - 2y}{2x - 4y}$

$\frac{dy}{dx} = \frac{-1 - 2(-4) - 2(1)}{} \quad -1 + 8 - 2$

On each side of the diagram, the numbers in the circles subtract to give the number in the square.

Fill in the missing values.

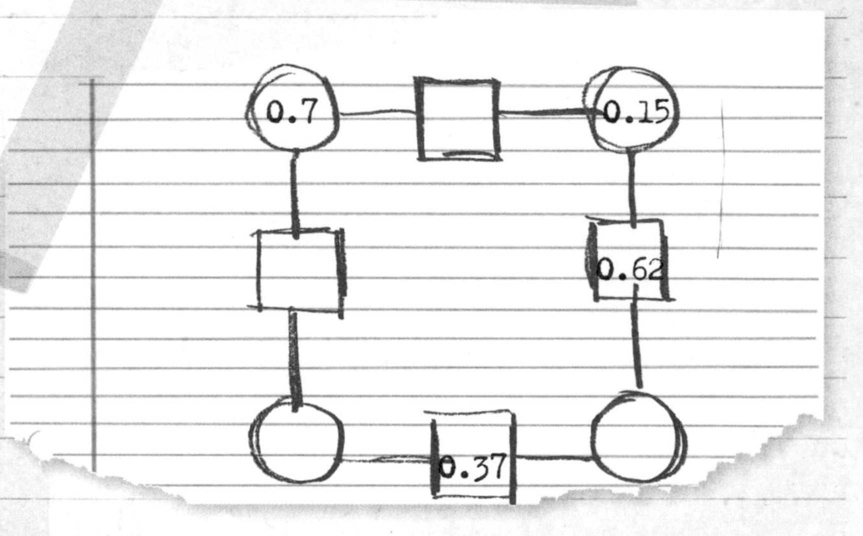

Three-quarters of a number is 54.

What is the number?

answers p122/3

Oranges are packed in trays of 24.
The trays are packed in boxes.
Each box contains 192 oranges.

How many trays are packed into a box?

"All the effects of nature are only mathematical results of a small number of immutable laws."

[Pierre-Simon Laplace]

Write the fractions in order of size, starting with the lowest.

$\frac{3}{4}$ $\frac{2}{3}$ $\frac{1}{2}$ $\frac{5}{6}$ $\frac{7}{12}$

6

answers p122/3

The ratio of male to female students in a class is 4:5

If there are 36 students in the class, how many are female?

[answers p122/3]

Here are four digit cards.

Use two of the digit cards to create a fraction to make each statement true.

1)

$$\frac{?}{?} = \frac{6}{15}$$

2)

$$\frac{?}{?} < \frac{1}{3}$$

"Truth is ever to be found in the simplicity, and not in the multiplicity and confusion of things."

[Isaac Newton]

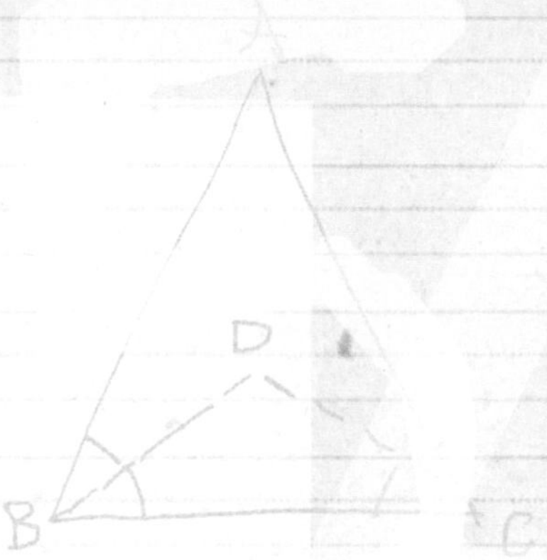

I am a common multiple of 4 and 6.

I am less than 50.

I am a square number.

What number am I?

page 46/147
145 x 23 = 3,335
Use this information to work out the answer to:
146 x 230
answers p122/3

Write the same digit in each box to make the multiplication correct.

□□ x □ = 704

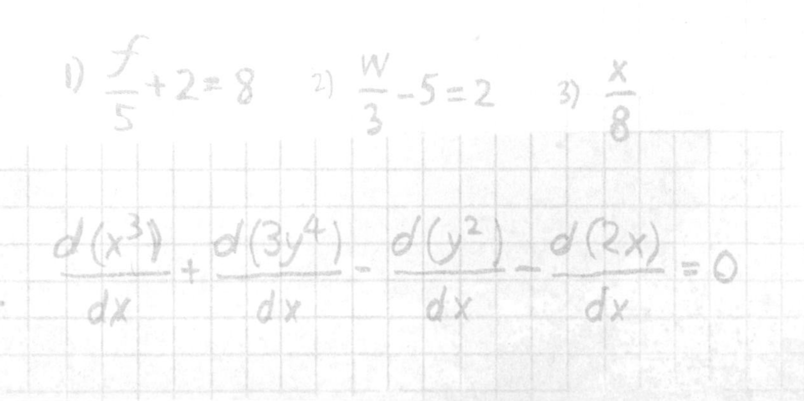

answers p122/3

Here is a spinner divided into eight equal sections.

What is the probability of scoring:

1) 1

2) 5

3) 3

4) 4

X and Y represent different whole numbers.

$X + Y = 540$

Y is 48 more than X

Calculate the values of X and Y.

In a survey, three-eighths of those asked a particular question answered YES.
The rest answered NO.

If 46 more people answered NO than YES, how many people were in the survey?

"The laws of nature are but the mathematical thoughts of God."

[Euclid]

answers p122/3

Work out the value of each shape.

$\square - \triangle = 5$

$\bigcirc + \bigcirc = \square \times \triangle$

$\square = 11$

answers p122/3

A bag contains six blue balls, two red balls, and eight white balls.

What is the probability of picking the following balls out of the bag?

Give your answers as fractions in the lowest terms.

1) A blue ball
2) A white ball
3) A red ball

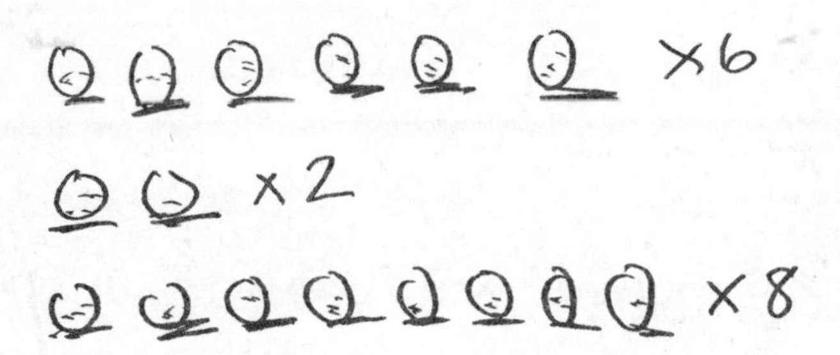

A small apple costs 12 cents and a large apple costs 25 cents.

Bob and Bill each have $5 to spend on as many apples as possible.

Bob buys only small apples. Bill buys only large apples.

How many more apples does Bob have than Bill?

answers p122/3

I think of a number.

I multiply the number by 5 then subtract 80 from the result.

My answer is the same as the number I started with.

Which number was I thinking of?

"Mathematics is like checkers in being suitable for the young, not too difficult, amusing, and without peril to the state."

[Plato]

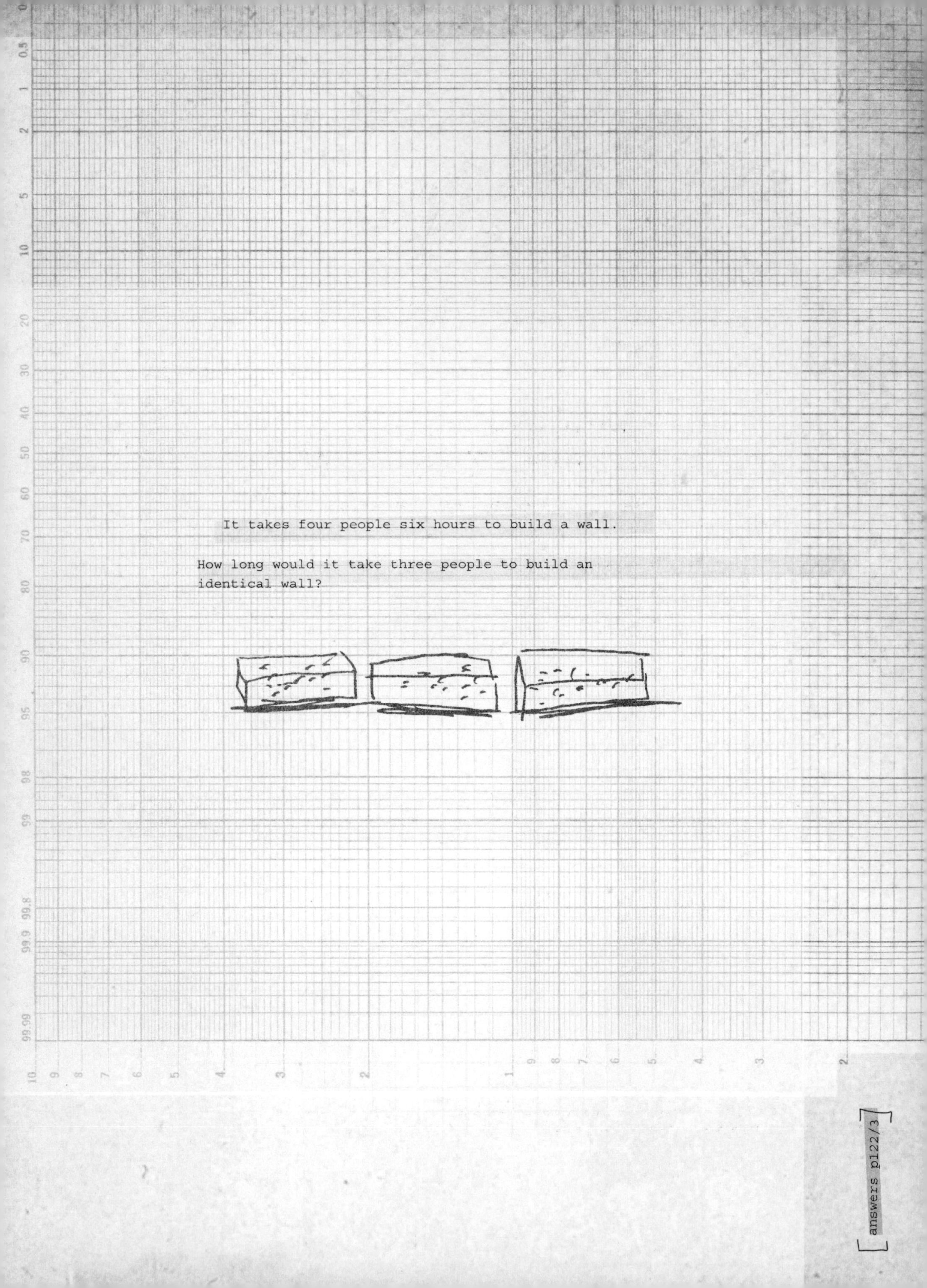

It takes four people six hours to build a wall.

How long would it take three people to build an identical wall?

[answers p122/3]

page 56/57

The numbers in each row, column, and diagonal add up to 34.

All the numbers from 1 to 16 appear, but only once each.

Fill in the missing numbers.

16		6	
	4		10
	14		8
2		12	

Write a single digit in each box to complete the calculation.

$\square 34 \times 3\square = 26{,}424$

"It is not of the essence of mathematics to be occupied with the ideas of number and quantity."

[George Boole]

I think of a number.

I add one-half of the number to one quarter of the number to give the answer 39.

What was my number?

answers p122/3

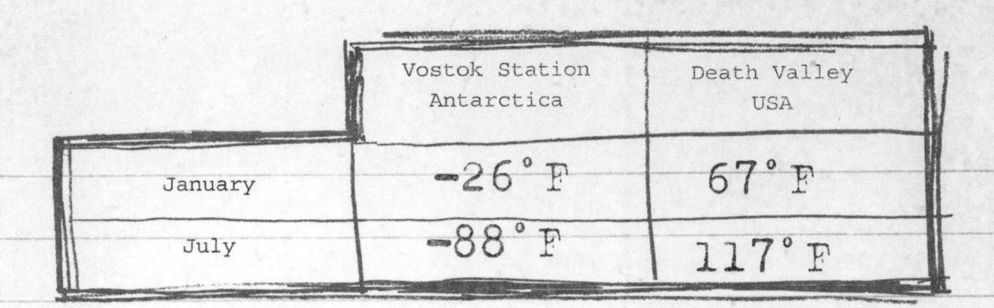

	Vostok Station Antarctica	Death Valley USA
January	-26° F	67° F
July	-88° F	117° F

The average temperatures for January and July are given for two different places.

1) What is the difference between the January and July temperatures for Vostok?
2) What is the difference between the July temperatures for the two places?

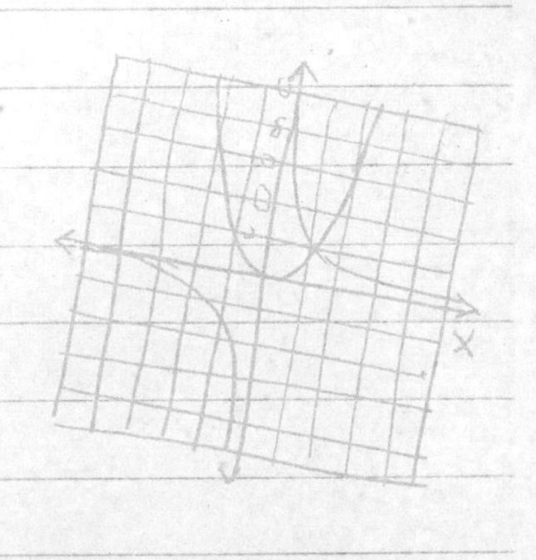

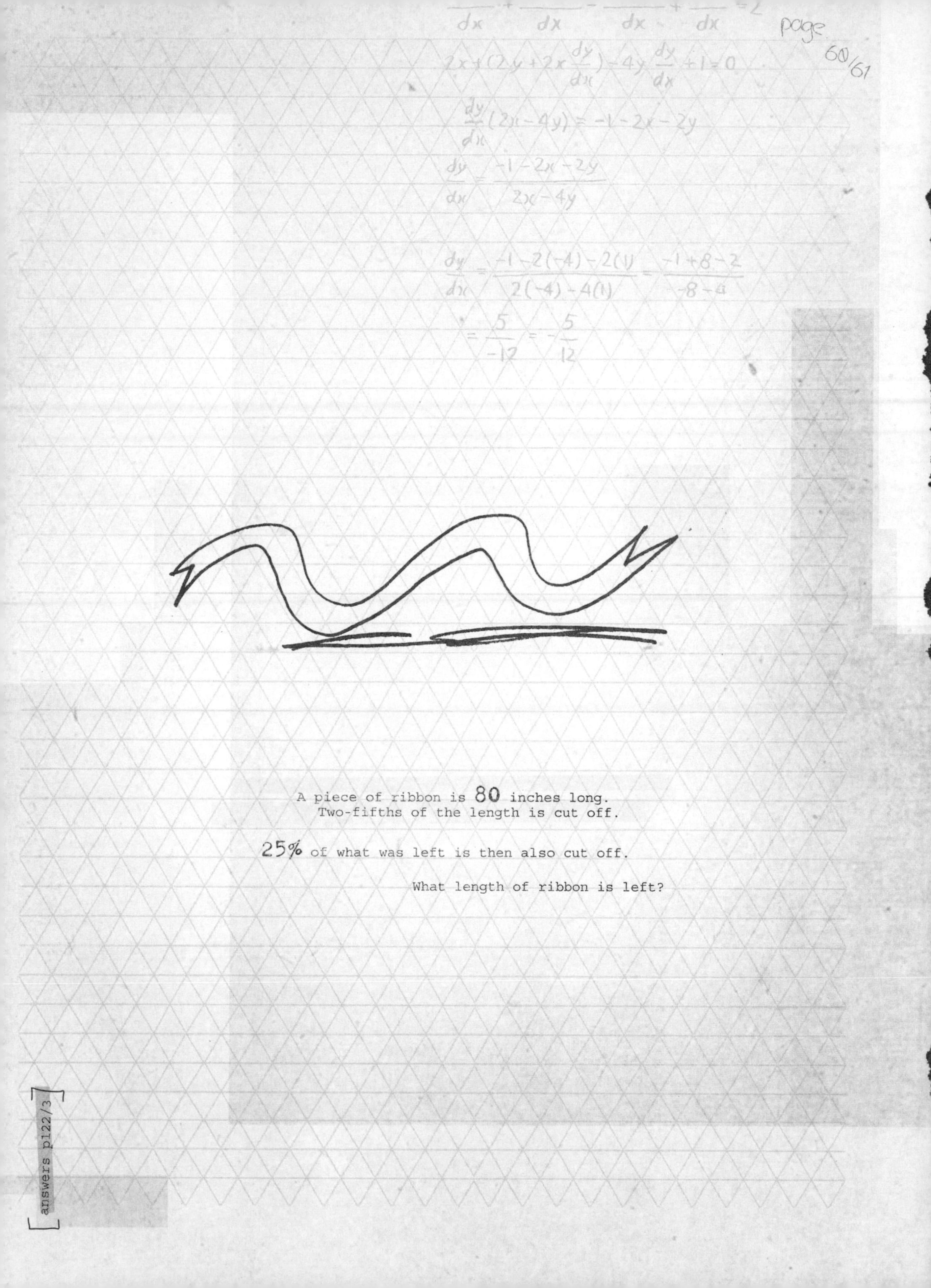
page 60/61
A piece of ribbon is 80 inches long.
Two-fifths of the length is cut off.
25% of what was left is then also cut off.
What length of ribbon is left?
answers p122/3

An animal lover has five dogs and an even number of cats.
He has more cats than dogs.

The total number of animals is less than 13.

How many cats does he own?

The clock is running fast. It shows the time as 9:09.

The digital watch is running 17 minutes slow.

How many minutes fast is the clock?

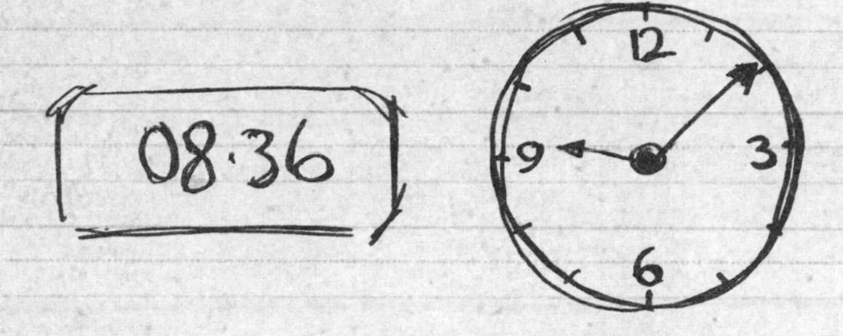

answers p122/3

"Wherever there is number, there is beauty."

[Proclus Diadochus]

There are 25 questions in a quiz.

A correct answer scores 2 points and an incorrect answer scores -1 point. No attempt to answer a question scores 0 points.

1) What are the maximum and minimum possible scores?

2) What score will a person get who attempts 21 questions and answers 13 of them correctly?

[answers p122/3]

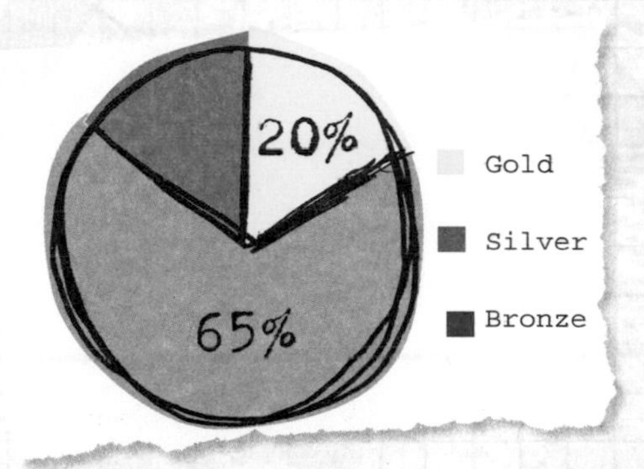

The pie chart shows the number of medals won by a country in the Olympic Games.

If 24 competitors won a gold medal, how many won a bronze medal?

Using only the four rules of numbers (addition, subtraction, multiplication, and division), and each number only once, make up a calculation that has the answer **403**.

25 7 6 2 9

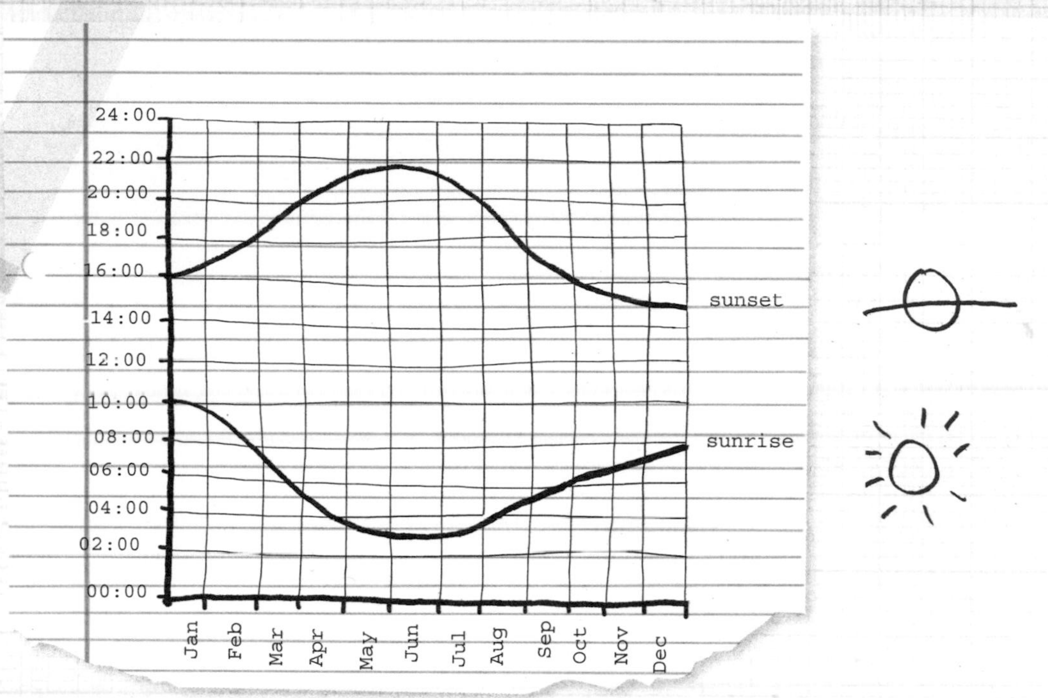

The graph shows the time that the sun rises and sets in a certain town.

In which months are there 12 hours between sunrise and sunset?

"The different branches of Arithmetic are Ambition, Distraction, Uglification, and Derision."

[Lewis Carroll]

answers p122/3

p, q, and r each represent a whole number. They add together to make 1,200.

$$p + q + r = 1{,}200$$

q is three times as big as r, and p is twice as big as r.

What are the values of p, q, and r?

When plotted on an x-y grid, the four points A, B, C, and D make a parallelogram.

Find the missing coordinates for points C and D.

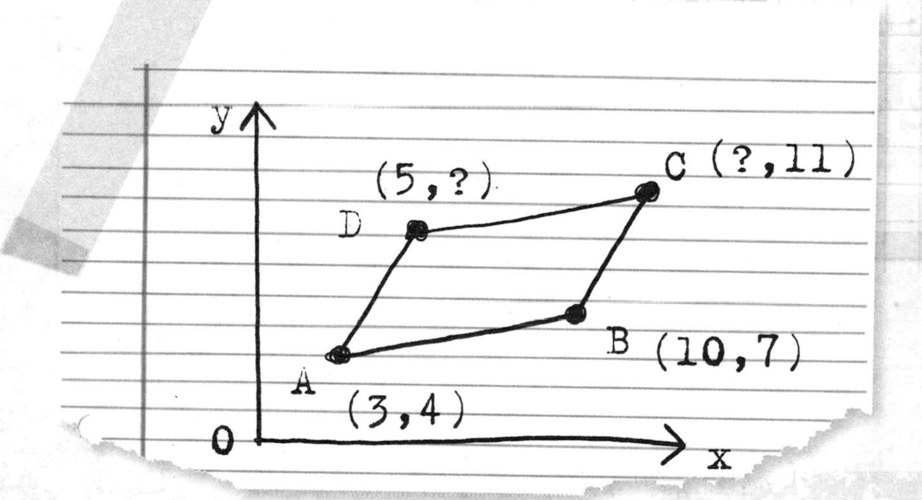

answers p122/3

"As far as the laws of mathematics refer to reality, they are not certain; and as far as they are certain, they do not refer to reality."

[Albert Einstein]

Write these in order of size, starting with the smallest.

$\frac{2}{5}$ 38% 0.35 $\frac{3}{8}$

The pie chart shows the favorite milkshake flavors for a group of students.
Sixty-five students' favorite flavor is strawberry.

How many students named vanilla as their favorite flavor?

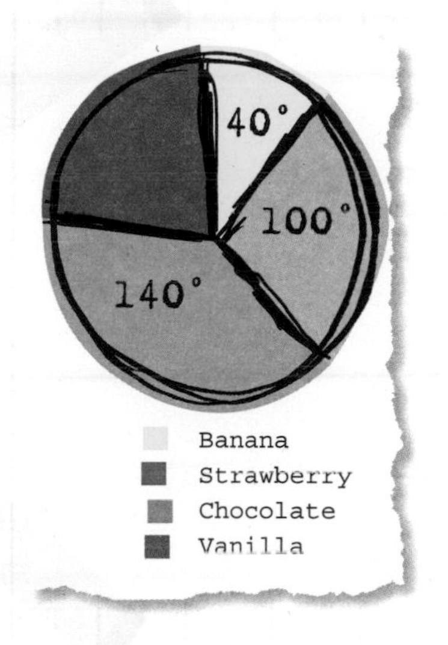

[answers p122/3]

An integer is a whole number, one that can be expressed without a decimal.

k stands for an integer greater than 15 and less than 20.

m stands for an integer greater than 5 and less than 10.

1) What is the smallest value of k x m?
2) What is the largest value of k - m?

[answers p124/5]

$$\frac{-2(-4)-2(1)}{(-4)-4(1)} = \frac{-1+8-2}{-8-4}$$

$$= -\frac{5}{12}$$

A group of five children has a mean age of **9** years.
Another child joins the group and the mean age
increases to **10** years.

How old is the child that joined the group?

Eggs can be bought in boxes of three different sizes.

Large box	24 eggs	$5.52
Medium box	18 eggs	$4.05
Small box	10 eggs	$2.35

Which box gives the best value for money?

"Either mathematics is too big for the human mind or the human mind is more than a machine."

[Kurt Gödel]

page 74/75

The squares of which two consecutive integers below **10** have a difference that is a square number?

answers p124/5

Complete each calculation using two numbers from the list.

−3 −6 4 −2 −8

□ ÷ □ = 2

□ x □ = −12

□ + □ = −4

□ − □ = 10

page 76/77

B Y a c A O A C

One of the four fraction calculations below has a different answer to the other three.

Which one is it?

$$\frac{3}{4} - \frac{1}{12}$$

$$\frac{9}{14} \div \frac{3}{7}$$

$$\frac{8}{9} \times \frac{3}{4}$$

$$\frac{1}{2} + \frac{1}{6}$$

[answers p124/5]

"Oh these mathematicians make me tired! When you ask them to work out a sum they take a piece of paper, cover it with rows of As, Bs, and Xs and Ys ... and then give you an answer that's all wrong!"
[Thomas Edison]

A board game uses a spinner and a normal six-sided dice.

1) List all the possible pairs of scores you could get with one roll of the dice and one spin of the spinner.

2) The two scores are added together. What is the probability of getting a score of 5?

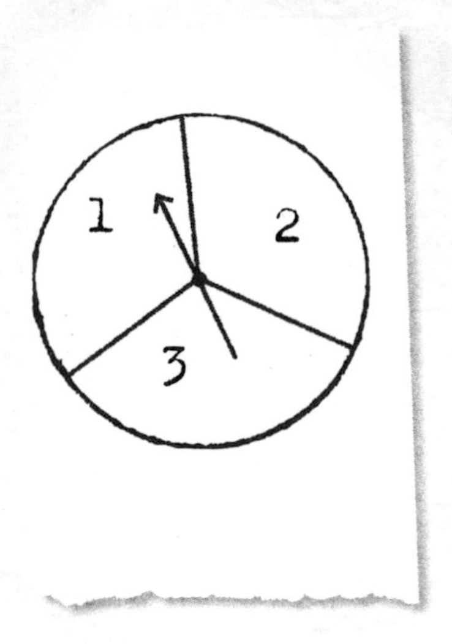

answers p124/5

page 78/79

A prime number only has two factors: 1 and the number itself.

Which three consecutive prime numbers have a product of 2,431?

$\square \times \square \times \square = 2{,}431$

A

Mathematical operations must be performed in the correct order: PEMDAS—Parentheses, Exponents, Multiply, Divide, Add, Subtract.

Work out the value of each sum.

1) $3 + 3 \times 3 + 3 \times 3^2$

2) $(3 + 3)^2 \div 3 + 3 \times 3$

A grocer has four bags of apples.

The first bag contains 19 apples.

The second bag contains 24 apples. The third bag contains 3 more apples than the fourth bag.

The mean number of apples is 22.

How many apples are in the third and fourth bags?

[answers p124/5]

Here are two boxes of buttons. The larger box contains more buttons than the smaller box.

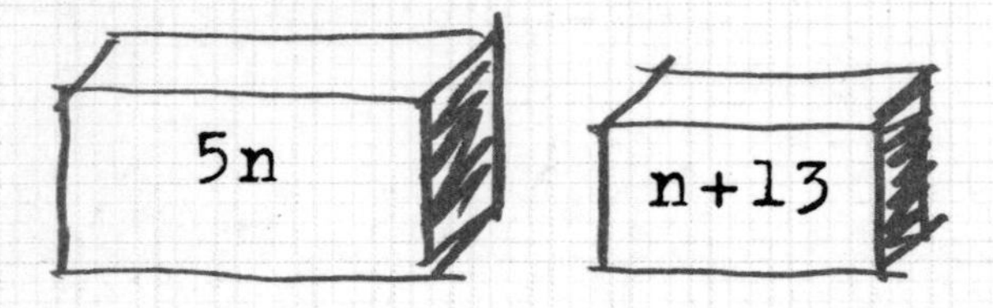

What is the smallest value of n?

"Arithmetic is being able to count up to twenty without taking off your shoes."

[Mickey Mouse]

page 82/83

When plotted on an x-y grid, the four points A, B, C, and D make a straight line. The four points are equally spaced along the line.

Find the coordinates for point A.

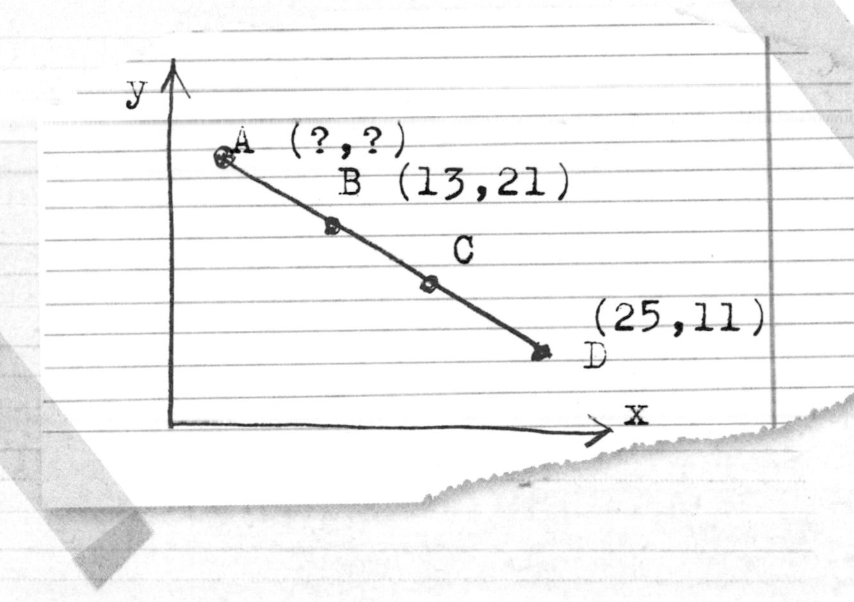

answers p124/5

For a set of one hundred numbers, the median is 80 and the mean is 85.

The highest number in the set is increased by 200.

What values are the median and mean now?

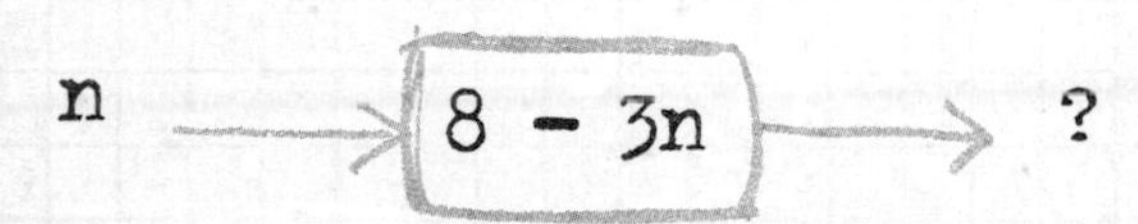

Work out the output from this function machine for each of these inputs:

1) $n = 2$

2) $n = 4$

3) $n = -3$

[answers p124/5]

7 10 13 16 19

The nth term of this sequence is $3n + 4$.

1) What are the 10th and 53rd terms?

2) Is the number 473 a term in the sequence?

"Mathematics consists of proving the most obvious thing in the least obvious way."

[George Polya]

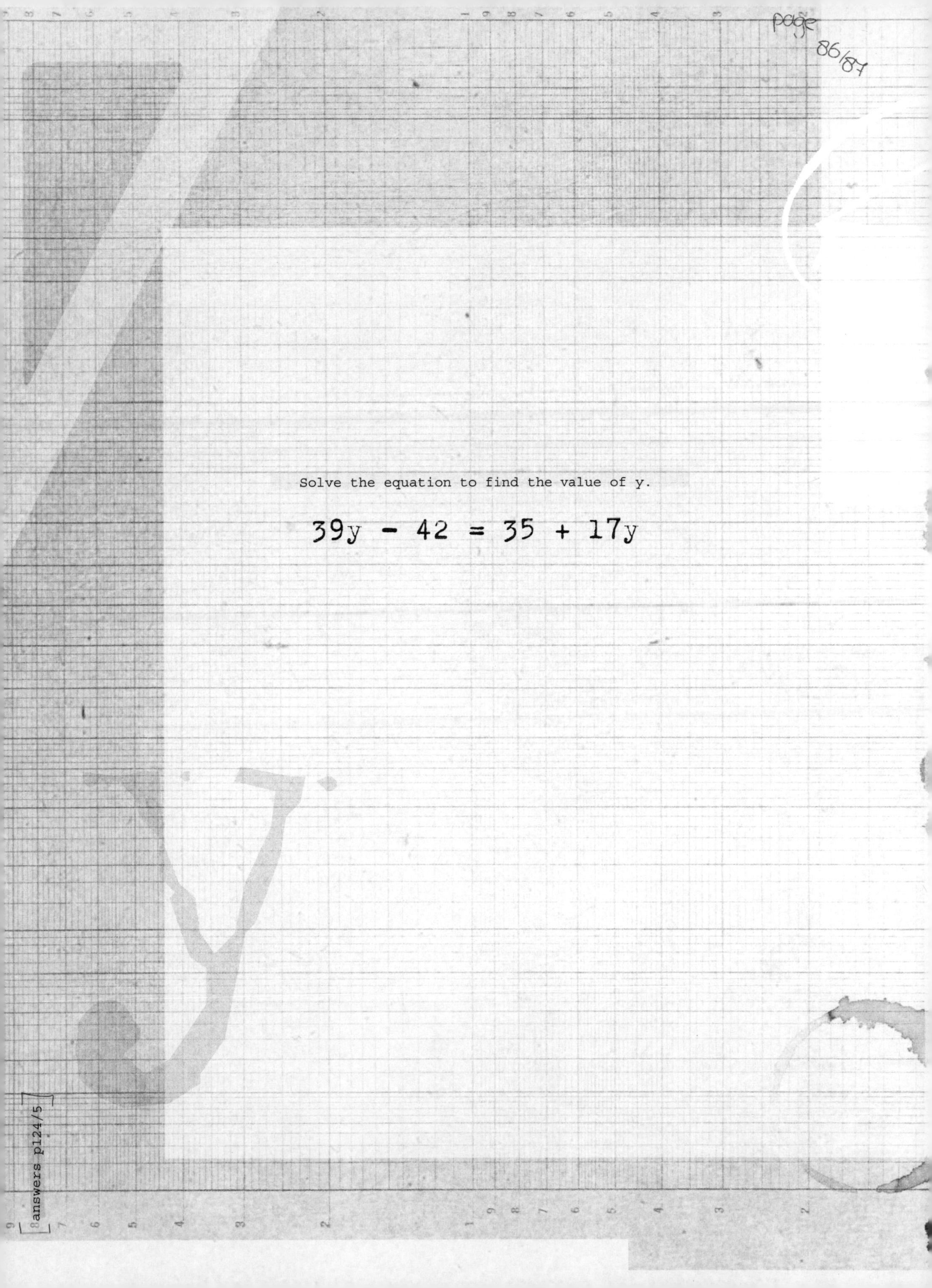

Solve the equation to find the value of y.

$$39y - 42 = 35 + 17y$$

answers p124/5

Work out four pairs of values for m and n that satisfy the equation.

$$m^n = 64$$

answers p124/5

The length of a rectangle, measured to the nearest inch, is 18 inches.

The width of a rectangle, measured to the nearest inch, is 10 inches.

1) What is the shortest length that the perimeter of the rectangle could be?

2) What is the longest length that the perimeter of the rectangle could be?

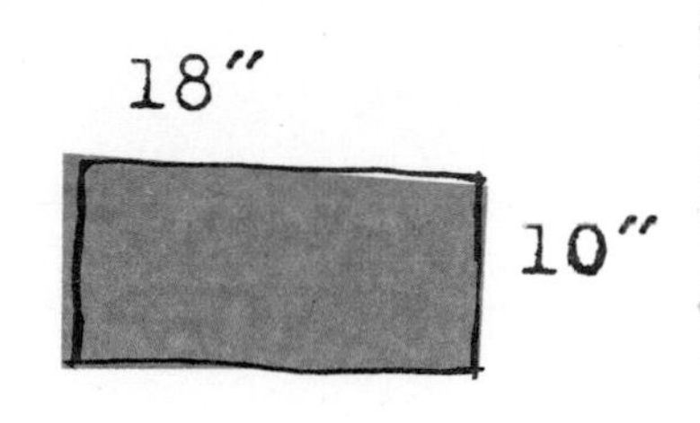

Here are two equations:

$$4x + y = 13$$
$$6x - y = 7$$

What are the values of x and y?

"The laws of mathematics are not merely human inventions or creations. They simply 'are'; they exist quite independently of the human intellect."

[M.C. Escher]

Make a the subject of each formula.

1) $a + b = c$

2) $2a + 1 = b$

3) $\frac{a}{b} = c$

4) $3 - a = c$

[answers p124/5]

Here are five algebraic statements.

$$3a + b = 15 \qquad 12a = 60 - 4b$$

$$b = 3a + 15 \qquad 6a + 2b = 30$$

$$b = 15 - 3a$$

Four of them are equivalent to each other, one is not.

Which is the odd one out?

[answers p124/5]

Here are two bags of coins.

There is the same number of coins in each bag.

1) What is the value of y?
2) How many coins are there in each bag?

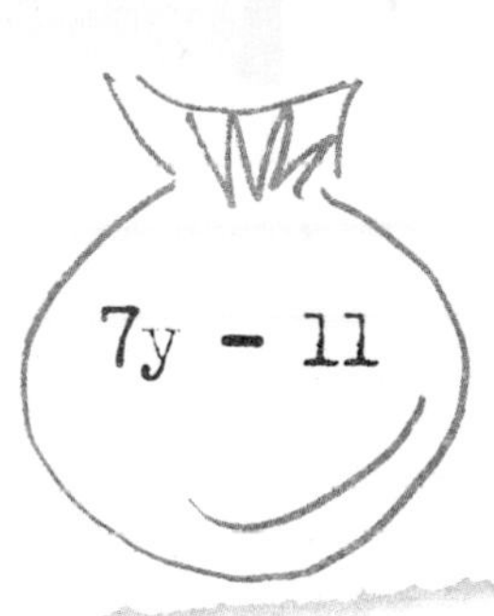

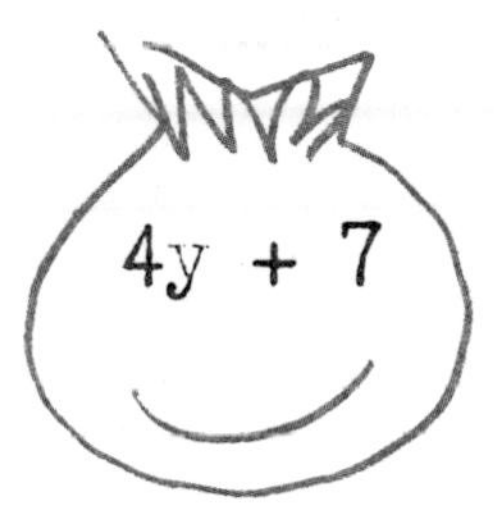

An amount is increased by 20% then decreased by 12%.

What is the equivalent overall percentage change?

"Mathematics is made of 50 percent formulas, 50 percent proofs, and 50 percent imagination."

[Henri Poincaré]

Find the value of x in the equation.

$$3^5 + 10^2 = 7^x$$

[answers p124/5]

Given that $a = 3$ and $b = 12$, work out the value of each expression.

1) $a + b$

2) ab

3) $\frac{b}{a}$

4) $(a + b)^2$

answers p124/5

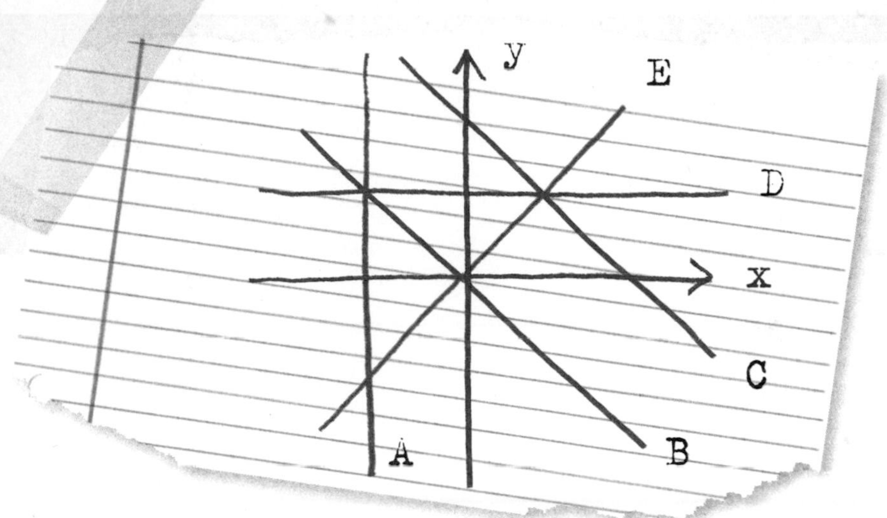

Here are five straight-line graphs.

Match each graph to the correct equation.

$x + y = 2$

$y - x = 0$

$y = 1$

$x = -y$

$x = -1$

If $P^2 = 10$, find the missing number in each equation.

1) $P^4 = ?$

2) $P^? = 10,000$

"The essence of mathematics is not to make simple things complicated, but to make complicated things simple."

[Stanley Gudder]

The nth term of the sequence 1, 4, 9, 16... is n^2.

What is the nth term for each of these sequences:

1) 0, 3, 8, 15...

2) 0, 1, 4, 9...

3) 2, 8, 18, 32 ...

[answers p124/5]

Solve the equation to find the value of x.

$$\frac{5\ (3x - 4)}{2x} = 7$$

What are the coordinates of A and B, the points where
the two graphs cross?

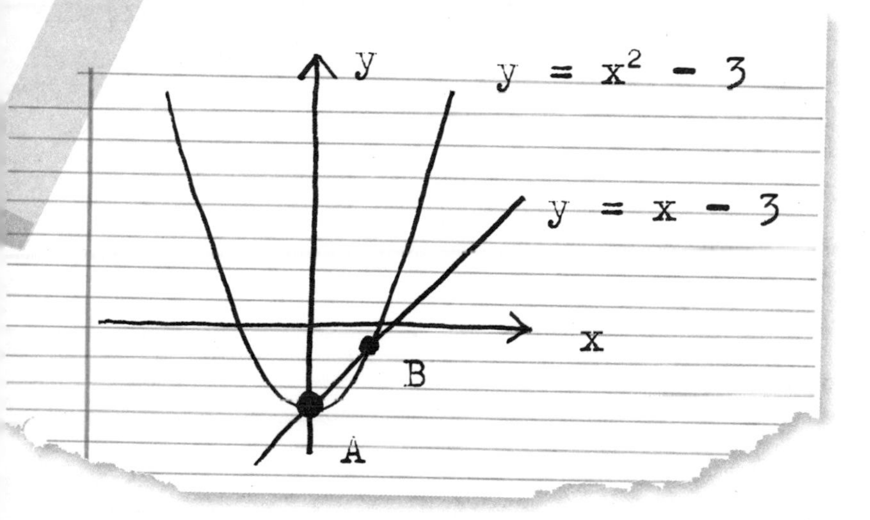

answers p126/7

A car lost 14% of its value after one year. It is now worth $20,167
What was the original value of the car?

answers p126/7

page 102/103

Solve the pair of simultaneous equations to find the values of a and b.

$$4a + 3b = 21$$

$$2a + b = 8$$

"When you can measure what you are talking about and express it in numbers, you know something about it."

[Lord William Thomson Kelvin]

Here are some powers of the number three:

$$3^{\frac{1}{2}} = \sqrt{3} \qquad 3^2 = 9 \qquad 3^{-1} = \frac{1}{3}$$

Write each of these in the form 3^n:

1) $\frac{1}{\sqrt{3}}$

2) $3\sqrt{3}$

3) $\frac{1}{9}$

4) 81

page 104/105

Find the solutions to the equation.

$$4k^2 = 1,156$$

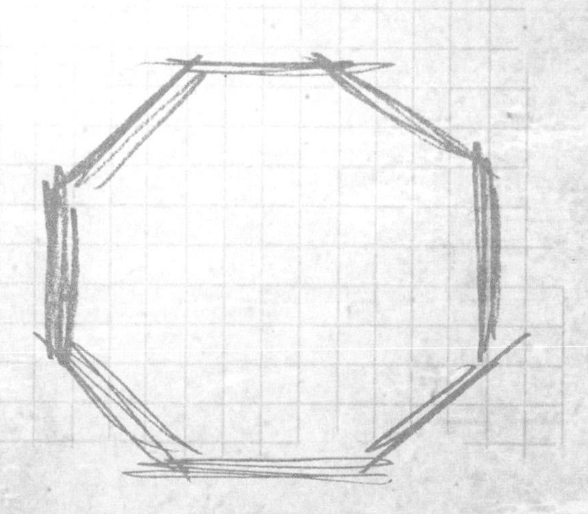

answers p126/7

"God does not care about our mathematical difficulties.
He integrates empirically."

[Albert Einstein]

The sum of two numbers is 80.

The difference of the two numbers is 26.

What are the two numbers?

answers p126/7

page 106/107

Some liquid is poured into a container.
A graph showing the change
in height with volume is shown.

1) Sketch the graph for this bottle.

2) Sketch the shape of the bottle to match this graph.

Find the matching pairs of equivalent algebraic expressions.

$2d \times 2d$

$d^3 \times d$

$d^2 + 3d^2$

$6d - 2d$

$d \times d \times d \times d$

$d + d + d + d$

Circle the three numbers that have the same numerical value.

3.4×10^{7}

34×10^{8}

$3.4 \times 10^{10} \times 10^{2}$

3.4×10^{9}

$3{,}400 \times 10^{6}$

0.34×10^{11}

answers p126/7

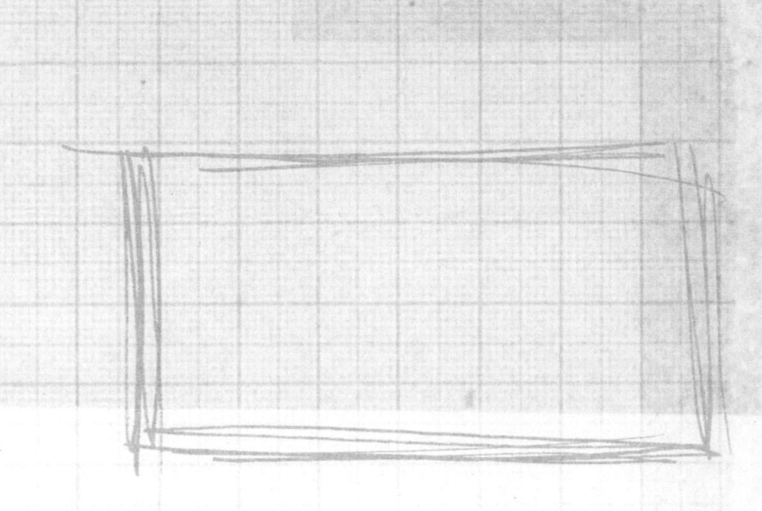

The rectangle has a length of (2a + b) and a width of (b - a).

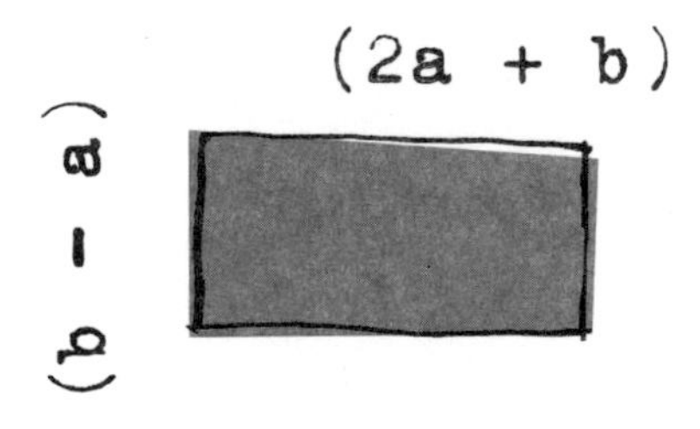

Find simplified expressions for both the perimeter and area of the rectangle.

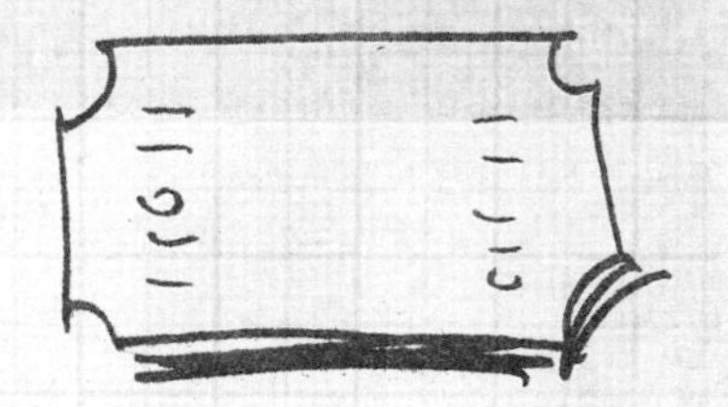

Two adult and three child
tickets for the movies
cost a total of $27.

The cost of movie tickets for
three adults and five
children is $42.50
altogether.

What is the cost of an adult
ticket and a child ticket?

"Every good mathematician is at least half a philosopher, and every good philosopher is at least half a mathematician."

[Friedrich Ludwig Gottlob Frege]

[answers p126/7]

A student needs to take an arithmetic and an algebra test.

The probability of passing the arithmetic test is 0.72.

The probability of passing the algebra test is 0.6.

What is the probability that the student passes only one of the two tests?

answers p126/7

y is inversely proportional to x.

When $x = 6$, $y = 12$.

What is the value of y when $x = 8$?

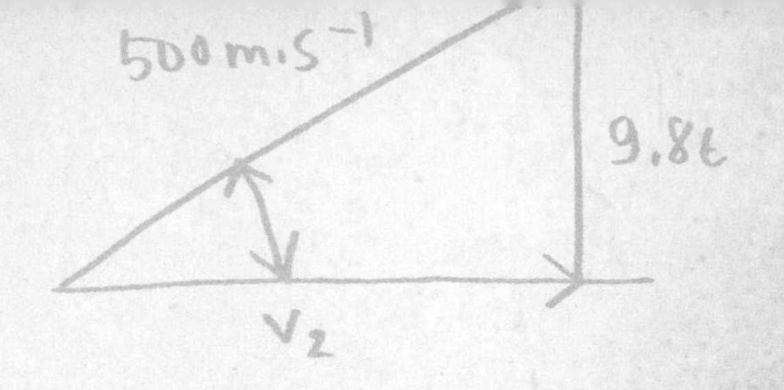

The formula for working out the final velocity of an object is given by:

$$v = u + at$$

where u is the initial velocity of the object, a is the acceleration, and t is the time taken.

Calculate t given that $u = 5$, $v = 11$, $a = 1.5$.

answers p126/7

An integer, n, satisfies these three

$n \leq 3$ $\quad$ $n > -2$ $\quad$ $-3 \leq n < 0$

What is the value of n?

Use the number line to help you.

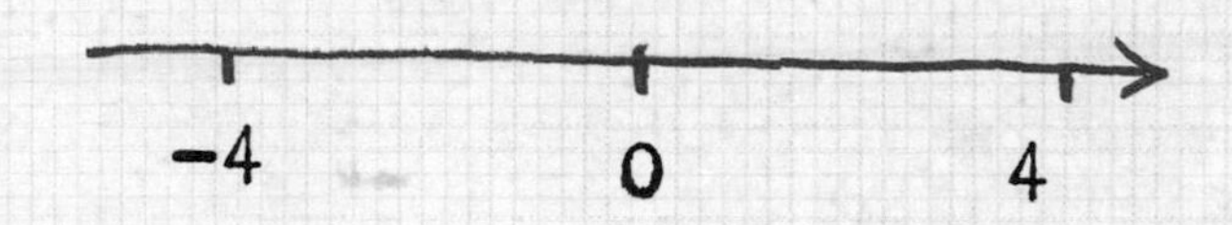

"A mathematician is a blind man in a dark room looking for a black cat which isn't there."

[Charles Darwin]

Find the pairs of index numbers that have the same value.

$27^{\frac{1}{3}}$ 3^{0} $9^{\frac{1}{2}}$ $3^{-4} \times 3$

3^{4} $\frac{1}{3^{3}}$ $\frac{3^{3}}{3^{2} \times 3}$ $(3^{2})^{2}$

answers p126/7

Factorize as fully as possible the following algebraic expression.

$$6a^2b^3c + 15a^3bc^2$$

Which one of these points does not lie on the line $3x + 2y = 6$?

(-3, 7.5) (0, 3) (-2, 6)

(1, -1.5) (4, -3)

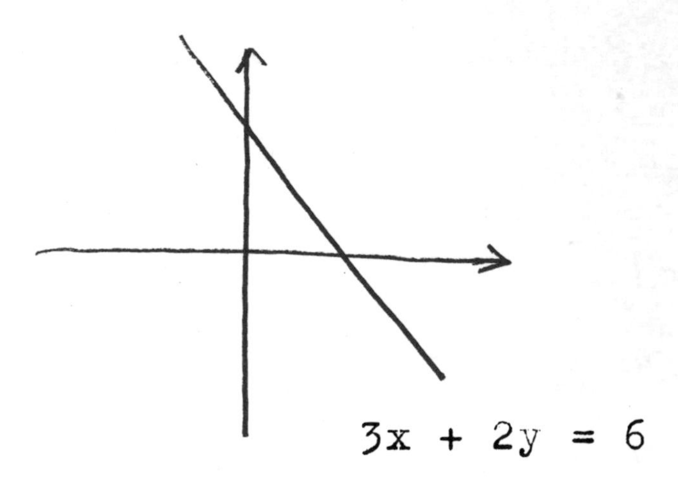

"You propound a complicated arithmetical problem: say cubing a number containing four digits. Give me a slate and half an hour's time, and I can produce a wrong answer."

[George Bernard Shaw]

The time taken, T, to complete one swing of a simple pendulum of length, L, is given by this formula:

$$T = 2\pi\sqrt{\frac{L}{g}}$$

Make L the subject of the formula.

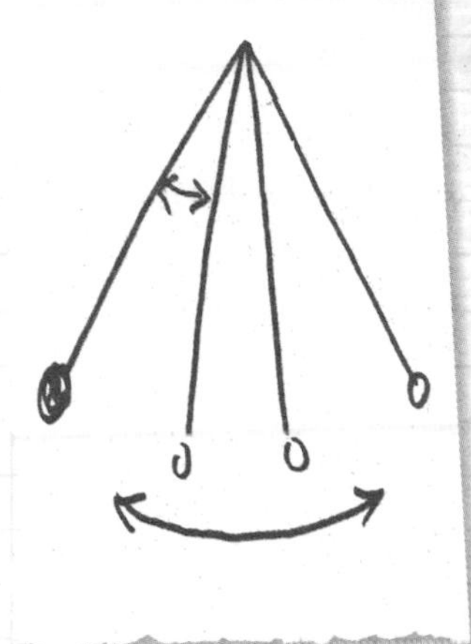

[answers p126/7]

p6

1) 6 and 5 (13 x 5 = 65)
2) 1 and 8 (18 x 2 = 36)

p7

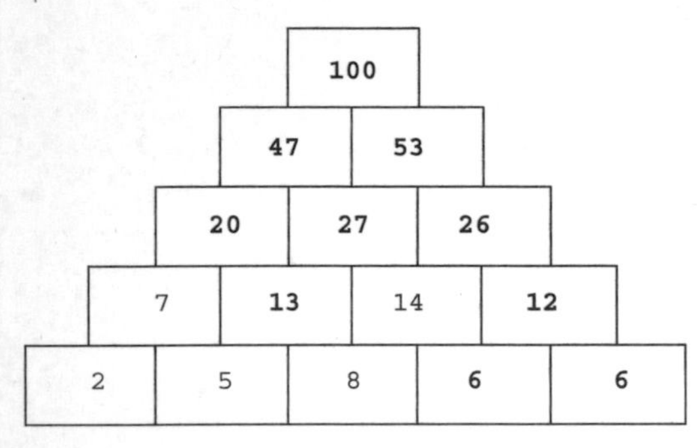

p8

1) Subtract 2 and add 5 (4 - 2 = 2; 2 + 5 = 7; 7 - 2 = 5; 5 + 5 = 10)
2) 8 and 13 (10 - 2 = 8; 8 + 5 = 13)

p9

Multiples of 6 only: 6, 12, 18, 30, 36, 42
Multiples of 8 only: 8, 16, 32, 40, 56
Multiples of both 6 and 8 (intersection):
24, 48

p10

5	X	**6**	=	30
X		X		
9	X	**7**	=	63
=		=		
45		**42**		

p11

1) 196 and 258 (196 + 258 = 454)
2) 171 and 304 (304 - 171 = 133)

p12

29, 53, and 68 (29 + 53 + 68 = 150)

p13

X	6	**7**	8	**4**
4	**24**	28	**32**	**16**
9	**54**	**63**	**72**	36
5	**30**	**35**	**40**	20
3	**18**	21	**24**	**12**

p14

92 and 65 or 95 and 62 (gives a total of 157)
62 and 59 (gives a difference of 3)

p15

1,652 (1,056 + 2,248 = 3,304;
3,304 ÷ 2 = 1,652)

p16

1 x 48, 2 x 24, 3 x 16, 4 x 12, and 6 x 8 (all equal 48)

p17

14 and 18 (14 + 18 = 32; 14 x 18 = 252)

p18

810 (45 x 18)

p19

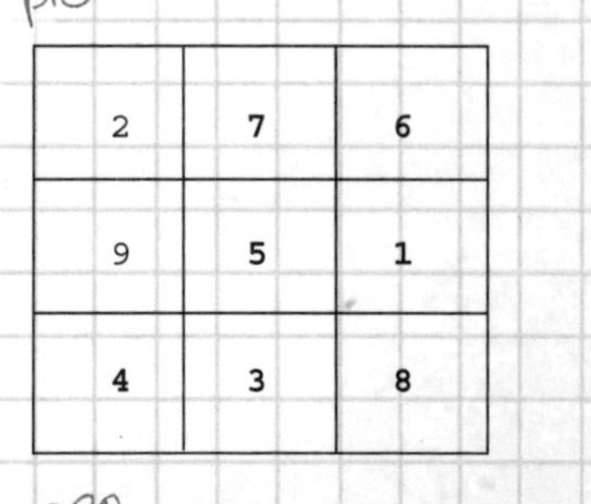

2	**7**	**6**
9	**5**	**1**
4	**3**	**8**

p20

A prime number is a number that can only be divided by 1 and itself.
7 x 13 = 91
2 x 3 x 11 = 66

p21

1,209,600 (60 x 60 x 24 x 14 = 1,209,600)

p22

5 (55 x 4 = 220; 1,100 ÷ 220 = 5)

p23

Car A: $1.35 (09:35 to 12:10 = 2 hours 35 minutes = $1.35)
Car B: $2.30 (07:50 to 11:55 = 4 hours 5 minutes = $2.30)

p24

23 (multiply the top triangle by the left triangle and add to the right triangle to get the center triangle, so 8 x 6 = 48, then 71 - 48 = 23)

p25

	2.5	0.52
0.3		
		0.5

p26

6 x 6 x 6 = 216
9 x 9 x 9 = 729

p27

1) Shade 8 parts (40% of 20 = 8)
2) Shade 4 parts (40% of 10 = 4)
3) Shade 2 parts (40% of 5 = 2)

p28

$^3/_5$, 0.6, and 60%

p29

```
   5 2 6
       6
x _______
 3,1 5 6
```

p30

15 is incorrect, it should be 17 (35 + 57 + 15 = 107 and 28 + 15 + 64 = 107 so the 15 must be wrong)

p31

67,500 people (15 + 40 = 55%, so 45% visited the rest of the year; 150,000 ÷ 100 x 45 = 67,500)

p32

6 x 6 = 36, 7 x 7 = 49, 8 x 8 = 64, 9 x 9 = 81, 10 x 10 = 100, 11 x 11 = 121, 12 x 12 = 144

p33

20 ÷ 8 = 2.5, so to find the right quantities you need to multiply each by 2.5, so:
15 ounces of flour
5 ounces of cornstarch
7.5 ounces of sugar
10 ounces of butter

p34

0.75 **1.5** 3 6 12 24 48 **96** **192**

p35

1) 56 kilometers (35 ÷ 5 x 8 = 56)
2) 30 miles (48 ÷ 8 x 5 = 30)

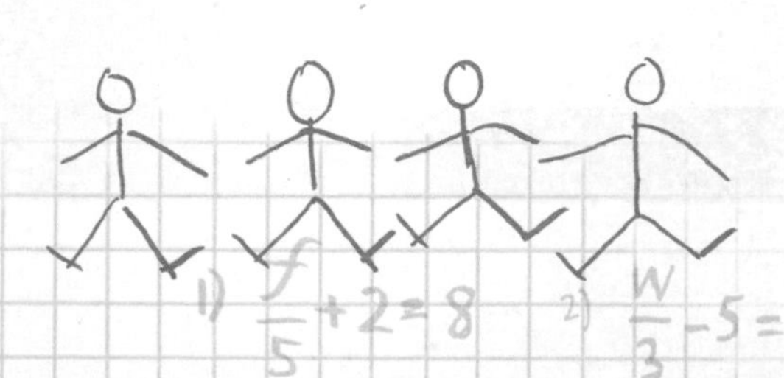

1) $\frac{f}{5}+2=8$ 2) $\frac{w}{3}-5=2$ 3) $\frac{x}{8}+3=12$

p36

1) 19 = 3 + 5 + 11
2) 23 = 5 + 7 + 11
3) 29 = 3 + 7 + 19

p37

58.906 (9.9 x 9.8 x 9.7 = 941.094;
1,000 - 941.094 = 58.906)

p38

$36^1/_2$, **31**, $25^1/_2$, **20**, $14^1/_2$, **9**, $3^1/_2$,
($36^1/_2$ - $25^1/_2$ = 11; 11 ÷ 2 = $5^1/_2$, so each number decreases by $5^1/_2$)

p39

13 miles (8.25 - 1.75 = 6.5 ÷ 0.50 = 13)

p40

Clockwise from top:
0.55 (0.7 - 0.15 = 0.55)
0.77 (0.15 + 0.62 = 0.77)
0.4 (0.77 - 0.37 = 0.4)
0.3 (0.7 - 0.4 = 0.3)

p41

72 (54 ÷ 3 x 4 = 72)

p42

8 (192 ÷ 24 = 8)

p43

Make all the fractions equivalent, with the same denominator:
$^9/_{12}$ $^8/_{12}$ $^6/_{12}$ $^{10}/_{12}$ $^7/_{12}$
Then put them in order:
$^1/_2$ $^7/_{12}$ $^2/_3$ $^3/_4$ $^5/_6$

p44

20 (for every 9 students, 5 are female, so 36 ÷ 9 = 4; 4 x 5 = 20)

p45

1) 2 and 5: $^2/_5$ = $^6/_{15}$
2) 2 and 7: $^2/_7$ < $^1/_3$

p46

36 (common multiples of both 4 and 6 that are less than 50 = 24 and 36, but 24 is not a square number)

p47

33,580 (145 x 23 = 3,335 + 23 = 3,358, then 146 x 230 = 3,358 x 10 = 33,580)

p48

88 x 8 = 704

p49

There are eight sections in the spinner, so the probabilities are:
1) 1 = $^1/_8$
2) 5 = $^4/_8$ or $^1/_2$
3) 3 = $^2/_8$ or $^1/_4$
4) 4 = $^0/_8$ or 0

p50

X = 246 (540 - 48 = 492 ÷ 2 = 246)
Y = 294 (246 + 48 = 294)

p51

184 (the difference in the number of people answering NO and YES = 46, so $^2/_8$ or $^1/_4$ ($^5/_8$ - $^3/_8$) of the people = 46; 46 x 4 = 184)

p52

Square = 11 (given)
Triangle = 6 (11 - 6 = 5)
Circle = 33 (11 x 6 = 66; 66 ÷ 2 = 33)

p53

There are 16 balls in total, so:
1) Blue = $^6/_{16}$ = $^3/_8$
2) White = $^8/_{16}$ = $^1/_2$
3) Red = $^2/_{16}$ = $^1/_8$

p54

21 (Bob can buy 41 small apples. Bill can buy 20 large apples. 41 - 20 = 21)

p55

The number is 20
The answer is best found by setting up an equation:
5 x number - 80 = number, or 5n - 80 = n
4n = 80; n = 20

p56

8 hours (4 ÷ 3 x 6 = 8)

p57

16	**9**	6	**3**
5	4	**15**	10
11	14	**1**	8
2	**7**	12	**13**

p58

7 and 6 (734 x 36 = 26,424)

p59

52 (39 = $^3/_4$ of the number, so 39 ÷ 3 x 4 = 52)

p60

1) 62°F (88 - 26 = 62)
2) 205°F (117 + 88 = 205)

p61

36 inches ($^2/_5$ of 80 = 32; 80 - 32 = 48;
25% of 48 = 12; 48 - 12 = 36)

p62

6 cats (13 - 5 = 8, BUT the total number is less than 13, so the number of cats is an even number between 5 and 8 = 6)

p63

16 minutes (the real time is 08:36 + 17 minutes = 08:53; the difference between 08:53 and 09:09 = 16 minutes)

p64

1) Maximum = 50 points (25 x 2); minimum = -25 (25 x -1)
2) 18 points (the person has given 13 correct answers and 8 incorrect answers, so (13 x 2) + (8 x -1) = 26 - 8 = 18)

p65

18 competitors (15% won bronze medals.
If 20% = 24, then the total competitors = 120.
120 ÷ 100 x 15 = 18)

p66

One possible calculation is:
7 + 9 = 16
16 x 25 = 400
6 ÷ 2 = 3
400 + 3 = 403

p67

March and September

p68

p = 400, q = 600, and r = 200
Make an equation:
q = 3r and p = 2r then 2r + 3r + r = 1,200;
6r = 1,200; r = 200, p = 400, q = 600

p69

The missing coordinate in C = 12 and in D = 8.
Since the difference in the x-coordinates of A and B is 7, so it is for C and D, hence the missing coordinate of C is 5 + 7 = 12.
Since the difference in y-coordinates of B and C is 4, so it is for A and D, hence the missing coordinate of D is 4 + 4 = 8.

p70

Turn all into decimals for easy comparison:
0.4, 0.38, 0.35, 0.375
Then put them in order:
0.35, $^3/_8$, 38%, $^2/_5$

p71

52 students (360 - 140 - 100 - 40 = 80, so vanilla = 80°; if 100° = 65 students then 80° = 52 students)

p72

1) 96 (16 x 6)
2) 13 (19 - 6)

p73

15 years old (9 x 5 = 45; 6 x 10 = 60; 60 - 45 = 15)

p74

Medium (each medium egg costs only 22.5 cents, 23.5 for small, and 23 for large)

p75

4 and 5 (difference of the squares of 4 and 5, 16 and 25, is 9, which is the square of 3)

p76

1) -6 ÷ -3 = 2
2) -3 x 4 = -12 (or 4 x -3)
3) -8 + 4 = -4 (or 4 + -8)
4) 4 - -6 = 10

p77

$^{9}/_{14} \div {}^{3}/_{7} = {}^{3}/_{2}$ (the rest have the answer $^{2}/_{3}$)

p78

1)
(1, 1), (1, 2), (1, 3)
(2, 1), (2, 2), (2, 3)
(3, 1), (3, 2), (3, 3)
(4, 1), (4, 2), (4, 3)
(5, 1), (5, 2), (5, 3)
(6, 1), (6, 2), (6, 3)
2) Probability of scoring 5 is $^{3}/_{18}$, or $^{1}/_{6}$

p79

11, 13, and 17 (11 x 13 x 17 = 2,431)

p80

1) 39
2) 21

p81

22 x 4 = 88; 88 - 19 - 24 = 45; the third and fourth bags contain 45 apples between them, so the third bag must contain 24 apples and the fourth bag 21 apples

p82

4 (5 x 4 = 20; 4 + 13 = 17)

p83

The coordinates for A are (7, 26)
Difference in x-coordinates of B and D is 12, making a difference of 6 for each pair of points, hence the x-coordinate of A is 13 - 6 = 7
Difference in y-coordinates of B and D is 10, making a difference of 5 for each pair of points, hence the y-coordinate of A is 21 + 5 = 26

p84

The median (the middle number in a list) remains the same at 80; the mean is 87 since the original total = 85 x 100 = 8,500, this increases to 8,700; 8,700 ÷ 100 = 87

p85

1) 8 - (3 x 2) = 2
2) 8 - (3 x 4) = -4
3) 8 - (3 x -3) = 17

p86

1) 10th term = 34 (10 x 3 + 4 = 34);
53rd term = 163 (53 x 3 + 4 = 163)
2) 3n + 4 = 473; 3n = 469; n = 156.33…; since n is not an integer, 473 is not a term in the sequence

p87

y = 3.5 (rearrange the equation so the variable is on one side only: 22y - 42 = 35; thus 22y = 77, so y = 77 ÷ 22 = 3.5)

p88

m = 64, n = 1
m = 8, n = 2
m = 4, n = 3
m = 2, n = 6
(Try out different whole numbers)

p89

1) 54 inches (shortest perimeter = 9.5 + 18.5 + 9.5 + 18.5 = 54)
2) 58 inches (longest perimeter = 10.5 + 18.5 + 10.5 + 18.5 = 58)

p90

x = 2 and y = 5
Use the following method:
Add the equations together to eliminate y
10x = 20; x = 2
Substitute x = 2 into the first equation
4 x 2 + y = 13
y = 13 - 8; y - 5

p91

1) $a = c - b$
2) $a = \frac{b - 1}{2}$
3) $a = bc$
4) $a = 3 - c$

p92

b = 3a + 15
6a + 2b = 30 divided throughout by 2 gives 3a + b = 15
12a = 60 - 4b rearranges to 12a + 4b = 60, which divided throughout by 4 gives 3a + b = 15
b = 15 - 3a rearranges to 3a + b = 15
b = 3a + 15 does not rearrange to 3a + b = 15 so is the odd one out

p93

1) y = 6 (7y - 11 = 4y + 7, so 7y = 4y + 18, so 3y = 18)
2) 31 coins in each bag (7 x 6 - 11 = 31)

p94

5.6% increase (20% increase = 120; 12% decrease of 120 = 14.4; 120 - 14.4 = 105.6; so total percentage change is 5.6% increase)

p95

x = 3 (3^5 = 243; 10^2 = 100, so 7^x = 343; 7 x 7 x 7 = 343, so x = 3)

p96

1) 3 + 12 = 15
2) 3 x 12 = 36
3) 12 ÷ 3 = 4
4) 15^2 = 225

p97

A: x = -1 [vertical line crossing x-axis at -1]
B: x = -y [line passing through (0,0) with gradient of -1]
C: x + y = 2 [line crossing x-axis at 2 with gradient of -1]
D: y = 1 [horizontal line crossing y-axis at 1]
E: y - x = 0 [line passing through (0,0) with gradient of 1]

p98

1) 100 ($p^4 = p^2 \times p^2$ = 10 x 10 = 100)
2) 8 (10,000 = 10 x 10 x 10 x 10 = $p^2 \times p^2 \times p^2 \times p^2 = p^8$)

p99

The sequence 1, 4, 9, 16,… is the sequence of square numbers, that is 1^2, 2^2, 3^2, 4^2, so:
1) $n^2 - 1$ (subtract 1 from each square ($n^2 - 1$): 1 - 1, 4 - 1, 9 - 1, 16 - 1 to give 0, 3, 8, 15)
2) $(n - 1)^2$ (square the numbers 0, 1, 2, 3 instead of 1, 2, 3, 4 that is $(n - 1)^2$: gives 0, 1, 4, 9)
3) $2n^2$ (double each square ($2n^2$): 2 x 1, 2 x 4, 2 x 9, 2 x 16 to give 2, 8, 18, 32)

p100

x = 20 (15x − 20 = 7 x 2x; 15x − 20 = 14x; 15x − 14x = 20 so x = 20)

p101

A (0, −3)
B (1, −2)
The graphs cross when $x^2 - 3 = x - 3$
Rearrange this equation to give $x^2 - x = 0$
Factorize to give $x(x - 1) = 0$
Gives solutions of x = 0 and x = 1
Substitute in y = x − 3 to get y = −3 and y = −2
Graphs cross at A (0, −3) and B (1, −2)

p102

$23,450 (20,167 x 100 ÷ 86 = 23,450)

p103

a = 1.5 and b = 5.
Cancel out the a by multiplying the second equation by −2 and adding the two equations together:
−4a − 2b = −16
3b − 2b = 21 − 16 so b = 5
Now substitute b = 5 into the original second equation to give 2a + 5 = 8; a = 1.5

p104

1) $3^{-1/2}$

$\frac{1}{\sqrt{3}} = \frac{1}{3^{1/2}} = 3^{-1/2}$

2) $3^{3/2}$

$3\sqrt{3} = 3^1 \times 3^{1/2} = 3^{3/2}$

3) 3^{-2}

$\frac{1}{9} = \frac{1}{3^2} = 3^{-2}$

4) 3^4

$81 = 9 \times 9 = 3^2 \times 3^2 = 3^4$

p105

k = 17 or −17 (k^2 = 1,156 ÷ 4 = 289; k = square root of 289 = 17 or −17)

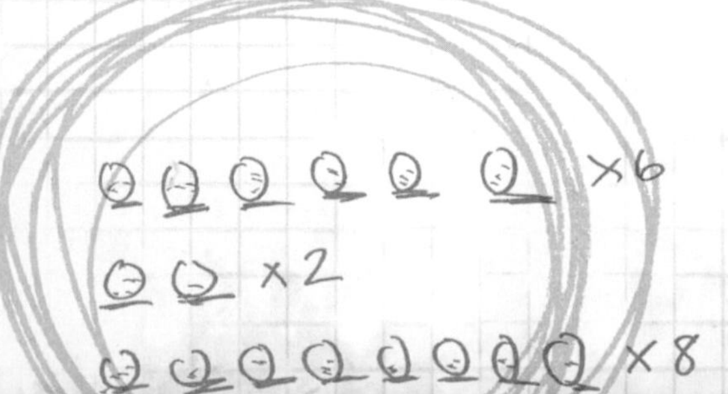

p106

27 and 53
Let the numbers be x and y and form two equations:
x + y = 80
x − y = 26
Add the two equations:
2x = 106; x = 53
Substitute in the first equation:
53 + y = 80; y = 27

p107

The straight-sided bottle fills at a linear rate from bottom to top.
The straight-sided bottle with a straight, narrow neck will fill at a slow linear rate in the main part of the bottle then at a faster linear rate for the neck, shown by a steeper straight line section on the graph.
The graph with three linear sections with each steeper than the previous part indicates a bottle with three straight sections, each one narrower than the previous one.

1)

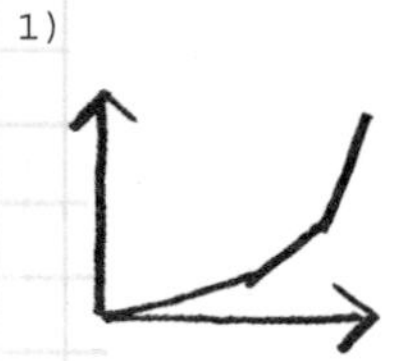

2)

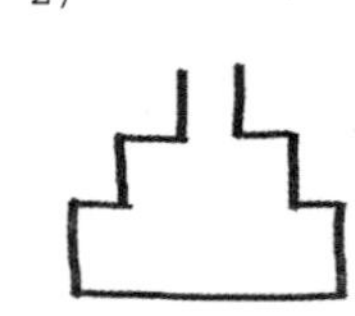

p108

2d x 2d and $d^2 + 3d^2$
d^3 x d and d x d x d x d
6d − 2d and d + d + d + d

p109

34 x 10^8, 3.4 x 10^9, and 3,400 x 10^6 (their numerical value is the same at 3,400,000,000)

p110

Perimeter = 2(2a + b) + 2(b − a) = 4a + 2b + 2b − 2a = **2a + 4b**
Area = (2a + b)(b − a) = 2ab − $2a^2$ + b^2 − ab = **b^2 + ab − $2a^2$**

p111

Adult ticket is $7.50 and child ticket is $4.00.
Let price of adult ticket be a and price of child ticket be c:
$2a + 3c = 27$
$3a + 5c = 42.5$
Multiply the first equation by 3 and the second by 2:
$6a + 9c = 81$
$6a + 10c = 85$
Subtract the new first equation from the new second equation:
$c = 4$
Substitute into the first equation:
$2a + 3 \times 4 = 27$
$2a = 15$; $a = 7.5$

p112

0.456
Probability of passing arithmetic test AND failing algebra test = $0.72 \times (1 - 0.6) = 0.288$
Probability of failing arithmetic test AND passing algebra test = $(1 - 0.72) \times 0.6 = 0.168$
Probability of passing only one test = $0.288 + 0.168 = 0.456$

p113

$y = 9$
Use the following method:
Substitute known values into $xy = k$: $6 \times 12 = 72$; $k = 72$
Substitute known values into $xy = k$: $8 \times y = 72$
So when $x = 8$, $y = 9$

p114

$t = 4$ ($11 = 5 + [1.5 \times t]$; $6 = 1.5 \times t$; $6 \div 1.5 = 4$ so $t = 4$)

p115

$n = -1$
Here are the possible integers for each inequality:

$n \leq 3$ 3, 2, 1, 0, -1, -2, -3, -4…
$n > -2$ -1, 0, 1, 2, 3, 4, 5…
$-3 \leq n < 0$ -3, -2, -1

The only number that appears in all three lists is -1

p116

$27^{1/3} = \sqrt[3]{27} = 3$ and $9^{1/2} = \sqrt{9} = 3$

$3^0 = 1$ and $\frac{3^3}{3^2 \times 3} = \frac{3^3}{3^3} = 1$

$3^4 = 81$ and $(3^2)^2 = 9^2 = 81$

$\frac{1}{3^2} = \frac{1}{27}$ and $3^{-4} \times 3 = \frac{1}{81} \times 3 = \frac{1}{27}$

p117

Split each term into factors, write the common factors outside the bracket and put the remaining factors from each term inside the bracket:
$6a^2b^3c + 15a^3bc^2 = \mathbf{3} \times 2 \times \mathbf{a} \times \mathbf{a} \times \mathbf{b} \times b \times b \times \mathbf{c} + \mathbf{3} \times 5 \times \mathbf{a} \times \mathbf{a} \times a \times \mathbf{b} \times \mathbf{c} \times c$
$= 3a^2bc(2b^2 + 5ac)$

p118

(1, -1.5)
Check by substituting each pair of values into the equation:
$3 \times (-3) + 2 \times 7.5 = 6$
$3 \times 0 + 2 \times 3 = 6$
$3 \times (-2) + 2 \times 6 = 6$
$3 \times 1 + 2 \times (-1.5) = 0$
$3 \times 4 + 2 \times (-3) = 6$

p119

$$T = 2\pi\sqrt{\frac{L}{g}}$$
$$\frac{T}{2\pi} = \sqrt{\frac{L}{g}}$$
$$\left(\frac{T}{2\pi}\right)^2 = \frac{L}{g}$$
$$\frac{L}{g} = \frac{T^2}{4\pi^2}$$
$$L = \frac{gT^2}{4\pi^2}$$

Thunder Bay Press
An imprint of the Baker & Taylor Publishing Group
10350 Barnes Canyon Road, San Diego, CA 92121
www.thunderbaybooks.com

All notations of errors or omissions should be addressed to Thunder Bay Press, Editorial Department, at the above address. All other correspondence (author inquiries, permissions) concerning the content of this book should be addressed to Paperwasp at the address below.

This book was conceived, designed, and produced by Paperwasp, an imprint of Balley Design Limited,
The Mews, 16 Wilbury Grove, Hove, East Sussex, BN3 3JQ, UK
www.paperwaspbooks.com.

Creative director: Simon Balley
Designer: Kevin Knight
Project editor: Sonya Newland
Illustrations: Kevin Knight

ISBN-13: 978-1-60710-441-4
ISBN-10: 1-60710-441-5

Printed in China.

1 2 3 4 5 16 15 14 13 12